UPGRADING CLUSTER
ENTERPRISES IN DEVELOI

Ashgate Economic Geography Series

Series Editors:
Michael Taylor, Peter Nijkamp, and Tom Leinbach

Innovative and stimulating, this quality series enlivens the field of economic geography and regional development, providing key volumes for academic use across a variety of disciplines. Exploring a broad range of interrelated topics, the series enhances our understanding of the dynamics of modern economies in developed and developing countries, as well as the dynamics of transition economies. It embraces both cutting edge research monographs and strongly themed edited volumes, thus offering significant added value to the field and to the individual topics addressed.

Other titles in the series:

The Moving Frontier
The Changing Geography of Production in Labour-Intensive Industries
Edited by Lois Labrianidis
ISBN: 978-0-7546-7448-1

Network Strategies in Europe
Developing the Future for Transport and ICT
Edited by Maria Giaoutzi and Peter Nijkamp
ISBN: 978-0-7546-7330-9

Tourism and Regional Development
New Pathways
Edited by Maria Giaoutzi and Peter Nijkamp
ISBN: 978-0-7546-4746-1

Alternative Currency Movements as a Challenge to Globalisation?
A Case Study of Manchester's Local Currency Networks
Peter North
ISBN: 978-0-7546-4591-7

The New European Rurality
Strategies for Small Firms
Edited by Teresa de Noronha Vaz, Eleanor J. Morgan and Peter Nijkamp
ISBN: 978-0-7546-4536-8

Upgrading Clusters and Small Enterprises in Developing Countries
Environmental, Labor, Innovation and Social Issues

Edited by

JOSE ANTONIO PUPPIM DE OLIVEIRA

Routledge
Taylor & Francis Group

LONDON AND NEW YORK

First published 2008 by Ashgate Publishing

2 Park Square, Milton Park, Abingdon, Oxon OX14 4RN
711 Third Avenue, New York, NY 10017, USA

Routledge is an imprint of the Taylor & Francis Group, an informa business

First issued in paperback 2016

British Library Cataloguing in Publication Data
Upgrading clusters and small enterprises : environmental,
  labor, innovation and social issues. - (Ashgate economic
  geography series)
  1. Industrial clusters - Asia 2. Industrial clusters -
  Latin America 3. Small business - Asia 4. Small business -
  Latin America 5. Social responsibility of business - Asia
  6. Social responsibility of business - Latin America
  I. Oliveira, Jose Antonio Puppim de, 1966-
  338.8'7

Library of Congress Cataloging-in-Publication Data
Upgrading clusters and small enterprises in developing countries : environmental, labor, innovation and social issues / [edited] by Jose Antonio Puppim de Oliveira.
    p. cm. -- (Ashgate economic geography series)
  Includes index.
  ISBN 978-0-7546-7297-5 (hbk) 1. Industrial clusters--Developing countries. 2. Small business--Developing countries. I. Oliveira, Jose Antonio Puppim de, 1966-

  HC59.72.D5U64 2008
  338.7--dc22

                                                        2008022337

ISBN 13: 978-0-7546-7297-5 (hbk)
ISBN 13: 978-1-138-25996-6 (pbk)

Contents

List of Figures

List of Tables

Preface and Acknowledgements

Many clusters of small and medium enterprises (SMEs) in Less Developed Countries (LDC) are counteracting the "race to the bottom" by becoming competitive while at the same time "socially upgrading": improving their innovation capacity, social impact, environmental and labor standards, and health-and-safety issues. There is significant literature on the competitiveness of clusters and SMEs, but little research about how and why competitive small firms in LDCs are socially upgrading. Issues such as global chains, Corporate Social Responsibility (CSR) and public policies have influenced the initiatives for cluster social upgrading. The objective of this book is to discuss a conceptual framework to understand what factors may lead clusters and SMEs to organize themselves to overcome obstacles to collective action for social upgrading.

The book is the result of several research initiatives related to clusters developed by a group of scholars in the last few years. Many of those activities were carried out within the research group on "Sustainable Development and Corporate Social Responsibility" (SD&CSR) of the Brazilian School of Public and Business Administration (EBAPE) at the Getulio Vargas Foundation (FGV), Rio de Janeiro, Brazil. The editor finished writing while at University of Santiago de Compostela, Spain. The group studied how and why social and environmental issues were introduced in the theory and practice of cluster development. During the meetings, the group found out that there were many publications on cluster upgrading, and environmental, innovation and labor issues in SMEs. However, the debates were not connected, even though they had a lot of links in common. One of the main objectives of the group was how to integrate the debates related to small firm/cluster upgrading in environmental, labor, social and innovation issues.

For this purpose, the editor organized two international workshops in Rio de Janeiro (Brazil) with leading researchers and practitioners to discuss the theme of cluster upgrading. The first was the "Workshop on the Institutional Challenges for Cluster Upgrading" on June 09, 2005. The second meeting was on 24 and 25 of August, 2006 and was called "Workshop on Upgrading Clusters: Experiences of Asia and Latin America". During the workshop, participants prepared papers and discussed practice-oriented research questions and emerging findings of their research projects and issues related to practice. Most of the chapters in this book resulted from the papers presented in the workshop.

Various organizations linked to the industry, applied research and academia had a chance to send their professionals to participate in the seminar to discuss the conceptual issues raised in the papers. The participants in the workshops came from the following Brazilian organizations: The Law School (GV Direito) and the Business School (EAESP) of Sao Paulo of the Getulio Vargas Foundation (FGV),

Research Institute of Applied Economics (IPEA), Catholic University of Rio de Janeiro (PUC-Rio), Federal Universities of Ceara (UFC) and Sao Carlos (UFScar), Technological Institute of Aeronautics (ITA), Center of Mineral Technology (CETEM), Government of the State of Bahia, Municipality of Petropolis, the Brazilian Service of Support for Micro and Small Enterprises (SEBRAE) and the Superintendency of the Manaus Free Trade Zone (SUFRAMA).

There were also participants from a number of international organizations including the Massachusetts Institute of Technology (MIT), Nagoya University, Institute of Social Studies (ISS), Center for Global Development (CGD), USAID, World Bank and International Labour Office (ILO).

However, the analysis and opinions in the book or their chapters reflect only the point of view of the authors and not of their organizations.

I would like to acknowledge several organizations that made this book possible, besides those that had participants in the workshops. Firstly, I thank the Brazilian School of Public and Business Administration (EBAPE) of the Getulio Vargas Foundation for providing the support and institutional environment for developing all the activities of the research group, specially its director, professor Bianor Cavalcanti, and its deputy dean for international relations, professor Eduardo Marques. I am also thankful to the Laboratorio do Territorio of the University of Santiago de Compostela, especially professor Rafael Crecente, and the European Commission through the Marie Curie Fellowship, which gave me the support to finish writing.

The Fulbright Foundation provided a grant through the Fulbright Senior Specialist Program to bring professor Judith Tendler (MIT) for a stay at EBAPE/ FGV in June 2005. The presence of ofessor Tendler was the motivation to organize the first workshop in 2005. This stay was also the seed to develop future activities of the SD&CSR research group, including the workshops and the theses.

The Japan Foundation sponsored the organization of the second workshop in 2006. The sponsorship allowed the presence of several Brazilian and international participants in the workshop in Rio. The Consulate of Japan in Rio de Janeiro was key to providing information about the existence of the grants of Japan Foundation.

The support of funds from the Ford Foundation Brazil was essential for having resources to edit this book.

Finally, I would like to express my special thanks to professor Judith Tendler of the Massachusetts Institute of Technology. Since the beginning, she supported and inspired the activities of the research group. Her stay at EBAPE and presence in the two workshops was fundamental for this book, even though she has no responsibilities over the quality of the chapters or the ideas and opinions expressed here.

Jose Antonio Puppim de Oliveira

Chapter 1

Introduction: Social Upgrading among Small Firms and Clusters

Jose Antonio Puppim de Oliveira[1]

1.1 Introduction: The Problems with Traditional Policies of Cluster Upgrading

In the last two decades, the cluster[2] concept has become popular both in academia and in practice. The cluster literature has grown significantly, as well as policies and tools that focus on creating and developing clusters.[3] As small and medium enterprises (SMEs) gained an important position in the development agenda as effective sources of jobs and income, policies to promote clusters came as a framework for augmenting the positive effects of small firms and optimizing resources to support them. Creating and supporting clusters would help small firms to overcome production and marketing obstacles they generally face, and allow them to compete with large firms and in sophisticated distant markets.

Thus, small firms, specially when agglomerated in clusters, could be important mechanisms for spurring innovation and dynamic economic development. Moreover, one of the main (and lovable) ideas behind those policies is that supporting small firms is also a good social policy, as they are weak but generate jobs and income, mainly for the poor. Many of the cluster policies in developing

---

1   Development Planning Unit (DPU), University College London (UCL), UK, and Laboratorio do Territorio (LaboraTe), the University of Santiago de Compostela (Spain). During the edition of this book he was at the Brazilian School of Public and Business Administration (EBAPE) of the Getulio Vargas Foundation (FGV), Rio de Janeiro (Brazil).

2   Cluster here is defined as an agglomeration of economic agents in the same economic sector operating close to each other geographically with some interaction or potential for interaction among them, and among them and supporting organizations, such as training centers, universities, government agencies, NGOs, unions and trade associations. Even though the focus of clusters is the dynamics of small firms, I also consider clusters economic agglomerations with the presence of medium and large firms, even multinationals. I do not distinguish the concept of cluster from similar others like Local Productive Arrangement (LPA) and Local Productive System (LPS), even though some authors in the following chapters make such distinction.

3   See the work of Schmitz (1995 and 2004), Schmitz and Musyck (1994), Schmitz and Nadvi (1999) and Altemburg and Meyer-Stamer (1999) and Altemburg and Eckhardt (2006).

countries and literature have assumed that supporting any kind of cluster/small firm—formal and informal, legal and illegal—leads to both local economic and social development, which is good for local communities. One typical example of those policies is microcredit for small, informal firms, now popular worlwide.

It is true that some policies lead to economic and social development, generating jobs and bringing people out of poverty and the reliance on social safety nets (where they exist). However, this development may not be sustainable over the long term, as SME policies may generate low-skill/low-paid informal jobs often under poor working conditions in businesses that produce poor quality products, do not pay taxes and do not respect environmental regulations. Benefits of those cluster initiatives can be wiped out suddenly by political turmoil, macroeconomic changes (e.g., changes in the value of the local currency) or international financial crises (Mead and Liedholm, 1998). Moreover, under globalization, competition from elsewhere or a weak position in a value chain can exacerbate the already poor business conditions making SMEs lose the little profit they had or reducing the already low salaries, leading to a "race to the bottom". They can still survive with subsidies, but as more price-competitive, high-quality imports begin to arrive, those limited economic and social benefits can disappear, sending petty entrepreneurs and their workers back into poverty or to reliance on social safety nets.

The policy studies and the literature on clusters have suggested that in order to avoid the fate described above ("race to the bottom"), clusters and their businesses should upgrade products, processes, functions and markets through continuous innovation (Humphrey and Schmitz, 2000). Upgrading would raise small businesses to another stage of economic dynamism, benefiting them and the broader dependent environment (workers, communities, local governments, etc.). However, there are two fundamental problems with the traditional idea of cluster upgrading which are analyzed in this book.

The first problem is that upgrading traditionally means improvements in the production or economic sphere of the clustered businesses. Upgrading would allow small firms to have better products or processes, reach better markets or reap more gains from their economic activity. However, this kind of upgrading pays little attention to the labor, social and environmental spheres. It assumes that by upgrading firms economically, their workers and the locality where they are based would benefit automatically (trickle-down effects). Economic upgrading would also mean more continuous innovation capacity or jobs for more qualified professionals. Indeed, many small firms in LDCs are able to learn, innovate, and employ qualified professionals, even in cutting-edge sectors (see Okada, 2004 and Okada chapter in this book). But this may not be true in all cases, as economically upgraded firms can also fire workers or squeeze their salaries and pollute more (see the chapter by Almeida in this book). Thus, traditional upgrading overlooks the social and environmental dimensions of the complete process of upgrading. A better upgrade would be what I call the *"Social Upgrading"*: an upgrade that would aim at a long-term development strategy based on formalized firms paying taxes; following environmental, labor, health and safety regulations and spurring social local development. There are many points in common between traditional upgrading and social upgrading[4], and sometimes they come together, but my point is that they are not the same thing.

The second fundamental problem of upgrading is the full understanding of how firms upgrade. It is not clear in the literature how and why firms would upgrade. The traditional cluster literature relies on the internal dynamics of the clusters for upgrading.[5] Through interaction among cluster members (firms, supporting organizations, etc.), firms would learn from each other. This would help them to innovate and develop new products, processes, functions and markets. However, many cases of upgrading include the interference of external actors, like actual or potential clients. Moreover, the literature does not distinguish the process of economic upgrading from social upgrading. It assumes that they are the same process, but they are actually different in many cases, as in the examples of this book (e.g., the important role of law enforcement officials).

Alongside the traditional cluster literature, the chapters in this book empirically examine three other basic frameworks which have been developed recently in the literature to explain ways that firms can socially upgrade: upgrading through markets, through ethical concerns (Corporate Social Responsibility—CSR) and through regulation. Although certain limitations remain, those frameworks can bring new insights to understand social upgrading.

First, upgrading through markets: some authors suggest that linking with global chains and markets—both generally more demanding in terms of quality— would bring the motivations and resources (technical and financial) to make firms choose to upgrade (Humphrey and Schmitz, 2002). There are also many cases of social upgrading through global chains, especially exporting markets in developed countries, which demand certain social and environmental standards. Many times those standards are regulations (like the directives of the European Union). However, the reality for most of the small firm clusters in developing countries gives little prospect for socially upgrading through global chains, as most of the firms are not linked to more demanding international or even national markets.

Second is upgrading through ethical and social principles. This is the basis of the framework of Corporate Social Responsibility (CSR). CSR involves the voluntary ethical actions to improve labor, environmental and social standards of firm stakeholders (internal and external). Small firms have implemented some CSR activities traditionally or through new approaches designed specifically for

---

4   In many parts of the book, the authors make a clear distinction between social upgrading and the traditional upgrading (sometimes also called economic upgrading or production upgrading). Other terms to differentiate other kinds of upgrading are also used, such as environmental upgrading, labor upgrading and technological upgrading. In some parts of the book the term upgrading (or upgrade) is used without differentiating the kind of upgrading, so that social and economic upgrading are used interchangeably.

5   See a criticism of that in Humphrey and Schmitz, 2002.

small enterprises (Vives, 2006). Nevertheless, even though CSR has the potential to improve social and environmental standards, especially for large firms and their chains, CSR has limited effects because it is still new and voluntary.

Third is upgrading through regulation. Social upgrading could be achieved by simply complying with the economic, labor or environmental laws, which are strict in many developing countries. However, regulatory compliance and enforcement are major problems in developing countries. There are political obstacles to making local governments support upgrading, and this may require further changes and even sacrifices on the part of the businesses. Many local firms prefer to make deals with local governments to extend subsidies and a lack of law enforcement, alleging they cannot compete and may close down if certain laws are applied. As cluster and SME policies in developing countries are a mix of social and economic development policy, many local politicians tend to accept this as a kind of social policy and a political token to keep the support of important constituencies, what is called the "devil's deal".[6]

The book explores the idea of social upgrading and the conditions under which it tends to take place. This chapter further discusses the idea of social upgrading and the ways to achieve it as well as how the other chapters contribute to that discussion. First, I will introduce the main debates on clusters. Second, I will explain the concept of social upgrading. Third, I will analyze the three main frameworks for social upgrading mentioned above. Finally, I will present a framework to advance the agenda of cluster social upgrading and the advantages of working with clusters for social upgrading in SMEs.

1.2 Debates on Clusters, Value Chains and the Social and Environmental Issues in SMEs

Even though there is no consensual definition of what a cluster is exactly (Martin and Sunley, 2003; Asheim et al., 2006), its principal idea is the agglomeration of firms in one sector of economic activity in a certain geographical space (Amorim, 1998; Cassiolato, et al., 2000; Schmitz, 1995). Besides the firms, clusters are also composed of other economic, social and political actors called supporting organizations, such as trade unions, distributors, export agents, governments and universities. Studies on clusters are currently in vogue and they are the object of study in several fields, such as development studies, business economics, regional sciences, political economy and geography (Porter, 1998; Markusen, 1996b; Piore and Sabel, 1984, Martin and Sunley, 2003).

---

6   Tendler (2002): "a kind of unspoken deal between politicians and their constituents, many small-firm owners, including those in the informal sector: 'If you vote for me, according to this exchange, I won't collect taxes from you; I won't make you comply with environmental, or labor regulations; and I will keep the police and inspectors from harassing you.'"

The concept of the cluster was particularly important to giving conceptual support to SME policies. In clusters, SMEs are able to overcome some of the obstacles they usually face when operating in isolation such as a lack of specialized skills; difficulty in accessing new technologies, inputs and services; problems in reaching markets and distribution channels; information, credit.

In order to overcome those obstacles, practical experience has shown that clustering can lead to several effects over time: division and specialization of labor; the emergence of a wide network of suppliers; the appearance of agents who sell to distant national and international markets; the emergence of specialized producer services; the development of a pool of specialized and skilled workers; and the formation of business associations. Moreover, because of the concentration of firms in one sector, many clusters also include specialized universities, as well as consulting, researching and training organizations, which can interact with the firms and create an atmosphere of intense knowledge production.

In general, the actors in a cluster have a certain degree of interaction among them exchanging information, goods and services. The dynamics of these interactions can bring greater advantages for firms that become part of a cluster rather than remain isolated. Many of the economic and organizational effects of clusters are result of the proximity and interaction among the economic actors and supporting organizations. These effects could be divided into two kinds. First, there are the external economies documented by Alfred Marshall in the nineteenth century (Marshall, 1890). They are the positive or negative unpaid, out-of-market-rules side effects (externalities) of the activity of one economic agent on other agents. There is also a second effect, the collective efficiency, which is the advantage to be gained by local external economies and joint action (Schmitz, 1995). This is a deliberate conscious act resulting from the collective action of different actors in the cluster. Those effects are important to explain cluster upgrading as collective action are necessary to help firms to overcome certain obstacles to upgrade.

Even though Marshall (1890) and others mentioned the advantages of spatial concentration of economic activities a long time ago, much of the modern work to explain the concept of clusters of small firms was developed in the Third Italy in the 1980s (Piore and Sabel, 1984). The region had a concentration of small and medium firms specialized in one specific sector that could compete in the global market with the gigantic "fordist" multinationals. The explanation for the competitiveness of those firms was the intense interaction that existed among them, as well as their flexibility in production (flexible specialization). The results of those studies gave major conceptual support to justify the significance of small and medium enterprises not only for social purposes (job generation, income distribution, etc.) but also as viable economic agents to compete in the global economy. Later on, the concept of clusters was expanded to include clusters among large and multinational firms, which maintained an innovative environment due to the cooperation as well as the intense competition among them (Porter, 1998).

Clusters have also been the subject of public policies in support of their development. Since the 1990s, government policies for creating and expanding clusters have been widespread, as one Brazilian public official said: "every region wants to create its own Silicon Valley". Many governments perceive the support for clusters as a way to generate dynamism in the economy and create foci of regional economic development. Thus, clusters have become both the unit of analysis and a framework for public action in economic development and industrial policies, especially in developing countries (Altemburg and Meyer-Stamer, 1999; Schmitz, 1995; Schmitz and Nadvi, 1999; Rabelotti and Schmitz, 1999; Meyer-Stamer, 1998). For example, the Brazilian Agency for the Support of Micro and Small Enterprises (SEBRAE) started to prioritize policies for supporting clusters instead of individual firms, which was the focus of SEBRAE's actions until the beginning of the 2000s.

The way governments promote clusters as a tool for industrial and regional development has been a source of debate in the literature. On the one hand, governments can be heavily interventionist to promote the development of clusters, such as providing incentives and market protection, together with infrastructure (Markusen, 1996a, 1996b). On the other hand, some authors argue that certain interventions can promote inefficiencies and hurt competitiveness in the long term (Porter, 1996). For them, governments should only provide infrastructure and support for education. The harsh competition among the firms in the same cluster (located next to each other) would be an incentive to innovate in order to make the clusters more economically efficient and competitive.

In the last years, the importance of the cluster concept has moved from the advantages of physical proximity, which generates collective benefits of the economies of scale and collective efficiency, to a more virtual proximity and interaction to generate knowledge and exchange of information (Nooteboom, 2006). Since the works of Marshall, economies of scale had been the main advantage for a firm being in a cluster. More recently, however, one of the main advantages has become the networking opportunities and knowledge generated among the different actors in a cluster. A firm in a cluster can take advantage of the immense opportunities for information exchange and knowledge generation through interaction with the different actors in the cluster. Over the years, the cluster concept has evolved beyond the physical geography of the economic agents and become more related to the social and economic interaction among the agents independently of their locations. There is no need for physical proximity among all actors in a cluster in a world with the Internet, cheap communications, easy capital flow, more efficient logistics and cheaper transport. Nowadays, suppliers, commercial agents, and supporting organizations can be far away from the firms without undermining the advantage of being close to each other.

The governance of clusters has been the subject of analyses in recent debates (Schmitz, 2004; Visser and Langen, 2006). Some of the inquiries examine what institutional and organizational mechanisms govern clusters and what are the roles of firms and supporting organizations, as well as external actors, in the governance of clusters. The concept of network (Powell, 1990) has been utilized as a framework of analysis to understand cluster governance (Atzema and Visser, 2007). Mechanisms of governance have been fundamental to spur innovation and upgrading in clusters (Schmitz, 2004; Nooteboom, 2006; see also the chapter by Ipiranga, Faria and Amorim in this book).

The literature on clusters had been characterized in the past as having a focus on the internal dynamics of the clusters to explain economic dynamics (Humphrey and Schmitz, 2002; Schmitz and Musyck, 1994; Rabellotti and Schmitz, 1999). Competitiveness, innovation and upgrading in a cluster were to be determined by the intensity of the relations among the different actors and their collective efficiency (Posthuma, 2004; Pyke et al., 1990). Also, there had been an emphasis on the need for intense cooperation among the different agents to lead to economic efficiency (Cassiolato and Szapiro, 2003). The cluster literature has given scarce attention to market relations (demand), and how markets influence changes in clusters (Tendler and Amorim, 1996), even though clusters are becoming more globalized as their economic relations expand to places far away from their activities. On the other hand, the literature on global value chains has been slow to accept their impact on clusters at the local level (Gereffi et al., 2003; Schmitz, 2005). Moreover, those two topics (clusters and value chain) have interacted very little, even though they analyze similar objects (Humphrey and Schmitz, 2002).

The analyses of external relations in the cluster literature have mostly been characterized by assuming an "ideal" market relationship, where there are many firms and buyers interacting through a competitive market (Cossentino et al., 1996). However, the relationship between firms and clients in clusters is much more complex, as the literature of global chains has shown. The arm's length, "perfect" market relationship is invalid in many cases. Clients can interact with suppliers in different forms in global chains, as shown in Table 1.1. Relations between client and supplier can go from a close market to a perfect market (arm's length) relationship, where the only relation is the commercial relation (sale) at a market price up to a complete verticalization (hierarchy) in the case of a multinational outsourcing production. In the middle, there is the network form of relation (strong interaction beyond the market, but no hierarchy) to a quasi hierarchy (client controls many aspect of the production). These types of relationships are important to our analysis of the governance in the case studies in this book.

There are several factors to explain upgrading in the labor, social and environmental standards, but there is no comprehensive literature to explain the conditions under which social and environmental improvements take place. Some economists make the argument of the (environmental) Kuznets curve: the improvement in those standards will come as economic welfare increases over certain levels (Anderson and Leal, 1992; Uhlig, 1992). Poor countries become more polluted and socially unequal as they grow economically; but when their per capita income reaches certain levels, environmental and social standards improve, as labor/environmental goods would become scarcer and people and markets would value more and be willing to pay for environmental goods and social equity.

Thus, economic objectives should come first for developing countries, as the others would come automatically with the increase in per capita income.

The cluster literature has also been slow to analyze noneconomic issues, such as environmental, labor, safety and health standards in clusters and other agglomerations (a few works have come out recently, see Damiani, 2003; Navdi and Waltring, 2004; Tewari and Pillai, 2004; Samstad and Pipkin, 2005; and Galli and Kucera, 2004). These issues may not be important in developed countries, where governments assure compliance of legal standards, but in many developing countries, even though some countries have solid regulatory frameworks, these are neglected because there is no political will or organizational capacity to enforce the regulations and implement public policies (Puppim de Oliveira, 2008, 2002). The general perception is that regulations, if enforced, are going to undermine competitiveness of firms and clusters. It is also politically difficult for authorities to "be tough" with small firms, once they are perceived to be weak and unable to comply with the laws (and if they tried to comply, they would be forced to close down). This assumes that the only advantage of those firms in developing countries is the advantage of price. However, many firms in LDCs have been able to be competitive and at the same time improve environmental, labor and health and safety standards. How this has happened is the object of analysis in this book.

Some studies have shown the importance of external actors to break cluster dynamics and enforce labor and environmental laws. In the case of Toritama in Northeastern Brazil, a new district attorney was key to enforcing existing environmental guidelines in a cluster of garment makers (See chapter by Almeida in this book). The linkages in the chain have also been important to explain environmental upgrading in clusters. Consumers and organized civil society including NGOs, especially in the developed countries, have pressed for socially and environmentally responsible products and processes. In the furniture sector, there is a growing demand for products with certified wood and certified management systems, such as ISO 14001 and SA8000, and demands for environmental quality regulations in the exporting market (see Chapter 3 by Puppim de Oliveira). In turn, global buyers have pushed down the chain for improvements in environmental standards. In India, the clusters in the leather industry have improved environmental standards to comply with demands from German clients (Tewari and Pillai, 2004), who also helped with the technical assistance to guide the environmental changes. In order to respond to these internal and external pressures, firms have cooperated with each other and with other actors in the clusters. They have helped one another to overcome the obstacles to environmental upgrading. This book tries to explain how such interactions have occurred and how the different links between clients and firms in certain clusters help to determine social and environmental changes within the clusters.

Table 1.1   Different linkages between clusters and global economy

Value chain linkage
Arm's-length market relationships: describes a relationship where there are potentially many buyers and sellers for equivalent products, even though particular buyers and sellers may engage in repeat transactions. This implies that the producer either makes a standard product or designs the product without reference to the needs of any particular customer. The customer is a "design taker". It also implies that there is no transaction-specific investment required by either party to the transaction.
Network relationships: occur when the supplier and buyer combine complementary competences. They may jointly design the product, using their different competences, and transaction-specific investment will be made. This type of relationship is particularly evident when both buyer and supplier are innovators, close to the technology or market frontiers, but it also arises when firms focus on their core competences and outsource important activities to suppliers.
Quasi-hierarchical relationships: occur when one party to the transaction (usually the buyer) exercises a high degree of control over the other. This often includes specifying the design (or the general specification) of what is to be produced and also process parameters such as quality systems, materials, etc. The introduction of monitoring and control procedures and the transmission of product design features requires transaction-specific investment.
Hierarchical relationships: occur, firstly, when the buyer takes ownership of the producers in the cluster or establishes its own companies within the cluster, or when firms in the cluster integrate forwards, establishing production or distribution facilities in other countries.

Source: Humphrey (2002) adapted from Humphrey and Schmitz (2000).

1.3 Going Beyond the Traditional Concept of Cluster Upgrading

The policy studies and the literature on clusters have suggested that in order to avoid the "race to the bottom", clusters and their firms should upgrade products, processes, functions and markets through continuous innovation. Upgrading means "making better products, making them more efficiently, or moving into more skilled activities" (Giuliani, Pietrobelli and Rabellotti, 2004). Recent cluster literature has attempted to better understand cluster upgrading (Schmitz, 2004). Upgrading generally involves a process of innovation and occurs gradually with marginal improvements in processes, products, functions and markets.[7]

---

7   Four kinds of upgrading are usually mentioned (i) *Process upgrading* is transforming inputs into outputs more efficiently by reorganizing the production system or introducing superior technology; (ii) *Product upgrading* is moving into more sophisticated product lines in terms of increased unit values; (iii) *Functional upgrading* is acquiring new, superior functions in the chain, such as design or marketing or abandoning existing low value-added functions to focus on higher value-added activities; and (iv) *Intersectoral*

However, the literature on cluster upgrading has focused on upgrading that brings increasing value added, and consequently better economic benefits for the firms in the cluster. Little attention has been given to the effects beyond the economic sphere, as other kinds of upgrading (which I call *Social Upgrading* here) such as upgrading environmental standards, labor practices, tax collection and health and safety standards, which are also important for local development. There is an overlapping of the lessons and concepts developed in the traditional cluster upgrading literature that can be used to understand social upgrading, as both try to avoid the "race to the bottom" (low road) path of business development. However there are some differences.

First, the traditional literature does not analyze the "trickle down" effects of cluster upgrading for local development. It assumes that upgrading a cluster will automatically bring positive impacts to local development. However, in the process of economic upgrading, many negative social and environmental impacts may occur. The upgrading process can lead to unemployment or an increase in the informal workforce, at the expense of the more abundant formal workforce. When unregulated, an upgrade may lead to an increase in production, as clusters gain markets, which in turn may lead to more environmental impacts (see chapter of Almeida in this book). Also, the conventional upgrading literature does not examine the effects on the general local/regional economy. Many benefits of cluster upgrading may not bring benefits to local development. This is particularly important in developing countries where law enforcement is weak and there is no secure bottom line on negative environmental and social effects of upgrading.

Second, the implementation of cluster upgrading may be different between the two kinds of upgrading (traditional, production or economic upgrading, and social upgrading). Social upgrading is directly related to process and product upgrading, and less to functional and inter-sectorial upgrading, even though the latter two types of upgrading may be important to motivate or allow social upgrading. Many of the benefits of an increasing quality of labor and environmental standards depend on improvements in process and product. Also, many actions for social upgrading may generate conflicts with the economic objectives of the enterprises. Firms may need to invest up front some resources to reach better labor and environmental standards, as well as pay the due taxes, which can threaten economic sustainability of the firms. In the exporting furniture firms, the high costs and investments needed were listed by the firms as the main obstacles to improve environmental standards (see the chapter by Puppim de Oliveira).

This book points to the limitations of the traditional upgrading in a globalizing economy. The chapters analyze empirically under what conditions and how firms have been socially upgrading. Three basic frameworks for social upgrading were found. The next section discusses those frameworks and how clusters and their firms can socially upgrade under each of them.

---

upgrading is applying the competence acquired in a particular function to move into a new sector. (Humphrey and Schmitz, 2000).

1.4 Frameworks for Socially Upgrading Firms

As discussed before, there is no general framework for clusters to upgrade socially. The traditional cluster literature has paid little attention to production or economic upgrading. This book identified three general frameworks for undertanding how clusters socially upgrade: linkage with global chains, CSR activities and better law enforcement. The linkage with global chains has shed some light on the cluster upgrading literature (Humphrey and Schmitz, 2002), and can also be an incentive to social upgrade. Corporate Social Responsibility (CSR) has emerged recently as voluntary ethical efforts by firms to social upgrade. Even though CSR influences only a minority of small and medium firms, it has the potential to push for leadership. Finally, the solution for cluster upgrading in developing countries may also come from an obvious action: make laws to be enforced. In the sections below, we discuss the opportunities and challenges for the three frameworks for socially upgrading.

1.4.1 Market Upgrading—Upgrading through Global Chains

One way to upgrade firms and clusters in developing countries is by linking them with global chains and markets, especially in more developed countries. Those markets and chains generally demand more in terms of quality, including labor/social/environmental quality. They also can bring the economic motivations (paying premiums for these qualities) and resources (technical and financial) to help firms upgrade.

Globalization can be an opportunity for clusters to go towards a "high road" dynamism under certain conditions. However, it also poses some challenges (see Rabellotti, 2001 and the chapter by Anne Posthuma in this book). The increasing global flow of goods, capital and information—results of advancements of trade agreements, telecommunications and logistics—has created new opportunities for firms to reach information and clients that were not reachable before (Humphrey, 2003). Nowadays, clusters and their firms can be better linked to distant or external markets and easily get different sources of information anywhere in the world through the Internet. Many consumers in more developed countries demand more socially and environmentally friendly products. Also, those countries set strict environmental quality regulations that affect all products coming to the country. Demands are pushed down the chain to firms outside those countries, including firms in developing countries. Those demands come as premiums for socially friendly products, or as rejection of socially unfriendly products.

Moreover, globalization has changed production and distribution chains. Chains are more complex and dispersed. One product can have parts from several different firms in various countries. Global Value Chain (GVC) analysis tries to understand how relations within a certain chain occur and how they can be governed (Gereffi, 1999). GVC is important to upgrading because GVC leaders—buyers or producers—can transfer knowledge or even financial resources to upgrade firms in developing countries. The participation of small firms from developing countries in those chains can be an opportunity to help them to socially upgrade, and consequently to have spillover effects for the communities.

There are two basic limitations in socially upgrading through global chains. The first is the governance of the chain—the relation between buyer and seller. It is fundamental to make the relationship benefit the firms in the developing country. Not all chains press for improvements in social and environmental standards. For example, buyers in developed countries can squeeze out small firms in developing countries to make them reduce their prices, caring little for the consequences, like lowered salaries and deteriorated working conditions. The (in)famous example of Nike in the 1980s is illustrative of this kind of behavior. The second limitation is the connection of firms with global chains. The conditions in most of the clusters of small firms in developing countries offer little opportunity to reach global chains of quality and more demanding international or even national markets. Especially in large developing markets like Brazil, India and China, most firms sell to local or national markets—generally less demanding in terms of social or environmental quality (see the chapter by Puppim de Oliveira in this book).

1.4.2 Upgrading through Corporate Social Responsibility (CSR)

The concept of Corporate Social Responsibility (CSR) can lead to another opportunity to upgrade clusters and their firms in developing countries. CSR is a voluntary initiative by firms to improve social/labor/environmental standards of their stakeholders because of ethical or market reasons. The modern debate of CSR was developed in the 1970s. It started in major corporations, but later was brought to small enterprises as well (Vives, 2006).

Even though there are some utilitarian reasons (social pressure or economic motivations) behind CSR initiatives such as gaining premium markets (described above in the GVC topic), ethical motivations exist as well. Firm managers and owners implement CSR initiatives because they think that doing more for the broader community is important. This can lead to improvements in working conditions and the social and natural environment around the firm. Moreover, one of the most important incentives for CSR in small firms is the influence of large firms in the chain. Large firms can demand better social/labor/environmental standards from their small suppliers, as part of their CSR commitments and policies. On the other hand, markets and political contexts are also changing, as new environmentally and socially concerned politicians, NGOs and consumers, especially in the high-end markets of developed countries, have been demanding changes in products and process, and pushing producers to socially upgrade (see chapter of Damiani in this book for the case of organic coffee in Mexico).Upgrading is the result of the efforts and the environment in which small businesses operate. The spread of the CSR framework in clusters can influence some firms to socially upgrade (because of the ethical or utilitarian motivations). Some governments and industrial unions have spread the idea of CSR and motivate some firms to do more than what actually is required (Cici, 2008). Even though CSR actions may have ethical motivations, they can also result in economic benefits, such as labor motivation and enhancing the public image of the company.

CSR has some limitations for upgrading small firms in developing countries. First, small firms may find it difficult to implement CSR policies, even if they are motivated, as they lack human, technical and financial resources. Second, CSR, as a voluntary inititiave, may reach only a limited number of firms, generally the leaders, which may not be where socially upgrading is necessary. Therefore, CSR is self-selected by the leaders in the top, but not the laggards in the bottom, which include generally most of the small firms and where changes are most needed. Third, the impact of CSR may be limited both in terms of socially upgrading the firm as well as the broader impact on local development. The upgrading led by CSR in a certain firm may be timid and benefit just a few stakeholders linked to the firm.

When CSR pressure comes from global chains, such as the need for social or environmental certification, the effects on small firms may be limited, as many of the firms are not linked to those chains. Certification generally pushes only the leaders to upgrade. For example, the demand for forest products certified by the Forest Stewardship Council (FSC) has grown exponentially in Brazil and throughout the world, but deforestation in the Amazon Basin has remained at high levels. Many of the drivers of deforestation are not linked to global chains or the CSR movements, such as firms that supply timber to local markets, illegal loggers or firms that deforest for other reasons (agriculture, cattle ranching, etc.).

1.4.3 Upgrading through Regulation—Enforcing the Laws

Another means of social upgrading are the enforcement of labor, environmental and tax laws. These laws exist in several developing countries, and are often very strict, but they are not properly enforced in many cases. For example, with regard to labor laws, more than half of the work force and almost two-thirds of small firms in the Brazilian economy were informal and operate illegally (IBGE, 1997 and 2003). The situation with environmental laws is very likely worse, as even many formal enterprises do not have the environmental license to operate. Regulatory reforms have focused on the high costs of an excessively strict and bureaucratic regulatory system. Reformers have claimed that dismantling the unbearable regulatory system would make it easier for firms to do business and generate economic development, but little attention has been paid on how certain clusters of SMEs are able to comply with law and be competitive.

Developing clusters by enforcing the law is particularly important in the case of developing countries, as globalization can push down the social, labor and environmental standards under certain conditions. The cluster analyses in developed countries[8] pay little attention to this as the labor/tax/environmental

---

8   See the works of Michael Porter (1996, 1998).

law is assumed to be automatically enforced. Definitively, the enforcement powers of developed countries are much stronger than in developing countries. Therefore, social upgrading is not an issue in the former. There is a bottom social/ environmental/labor line mandated by the law when companies deal with market forces: they cannot drive social standards below the regulated level without facing legal penalties. However, this is not the situation in developing countries where laws lack enforcement frequently. As clusters and small firms are exposed to markets, there is not a bottom line. Labor, social, and environmental conditions worsen in certain situations to enable small businesses to compete economically in the global market, and sometimes lead to environmental catastrophes and human rights abuses, such as child labor, etc.

Explanations of why laws are not enforced range from the lack of capacity of bureaucracies and weakness of the regulatory systems in many developing countries to the devil's deal (Tendler, 2002)—the local political situation that uses the threat of enforcement as a political bargaining tool to gain the support of an important constituency such as informal firms and workers.

Many cluster and small firm policies have paid little attention to social upgrading through regulation. Actually, many policies, influenced by the effects of the devil's deal, have given stronger incentives to remain in the informal sector. For example, small firms in the formal sector may find it difficult to access credit, and when they do, credit is expensive because of the high interest rates in many developing countries. On the other hand, microcredit is easier and cheaper for informal firms. The consequences of these policies are the creation of incentives to perpetuate low-quality firms and jobs, which may disappear in the long run, as well as poor labor, social and environmental conditions for the workers and local population as firms may not provide good working conditions, pay taxes or comply with environmental regulations.

Scarce research has been done to the devil's deal and how to break it. Indeed, there are some cases where firms were forced to comply with the "unbearable" regulation and became stronger and more competitive (see the chapter by Almeida in enforcement of environmental regulation). Therefore, if policymakers aim for local sustainable development, they may look at a long-term development strategy based on better regulatory enforcement and improved tax collection within the formal sector in order to begin to approach international standards.

1.5 Moving the Agenda on Clusters, Public Policies and Social Upgrading

Cluster development may lead to economic benefits for the participating firms and the broader local economy. One common solution to achieve that goal quickly is to relax the burdens on business operations, such as environmental, tax and labor regulations. Indeed, as commented before, many policymakers have concentrated much of their efforts on the "reform" of such regulations, as they see them as obstacles to spur economic development. However, many such benefits may be short termed or come at the expense of local environmental and social standards. Jobs generated may be low-skill and low-paid, and may not be maintained in the long run if other localities offer even lower production costs.

The literature on clusters in developing countries has given little thought to ways to upgrade because it has largely become self-encompassing, focusing on the internal dynamics of the clusters, emphasizing only their own economic benefits, and looking at the development of clusters as an end in itself. Upgrading, therefore, would be the automatic result of whatever interaction happens among the agents of a cluster. Limited conceptual thought has been done on cluster relations with external actors and how and why competitive clusters in LDCs are dealing with their social, environmental and labor standards, and health-and-safety issues. Also, there is little research on how the effects of cluster development trickle down to society in terms of local and regional development. These shortcomings in current cluster analysis and policy can lead to the development of clusters with short-term economic benefits, but limited long-term impacts for sustainable development. Unenhanced cluster development can only generate more low-paid informal jobs, cause significant environmental impacts (if regulations are not enforced), lead to an increase in accidents, and concentrate benefits in the hands of few.

Two fundamental questions related to public policy permeate the book in order to move the agenda of cluster, public policy and social upgrade: How and why do clusters socially upgrade? What should be the role of public policy in socially upgrading clusters and their firms?

The role of public policy is to support initiatives that look for long-term sustainable development, meaning to have a strong formal sector labor force working in firms that pay taxes and follow environmental, labor, health and safety regulations. The three frameworks described above could be the object of sound public policy as they can lead to social upgrading of clusters and their firms.

Meaningful public policies could begin to break the tradition of devil's deals and make law enforcement more effective. They could act in a kind of carrot-and-stick fashion, by both enforcing the law and providing the enabling conditions for firms to comply. For example, firms in the garment-making cluster of Toritama in Brazil started to comply with environmental laws when both the public attorney increased enforcement and they were given the resources (technology and finance) to adopt cleaner technologies (see the chapter by Almeida).

However, even though legal compliance would make the difference, there are limited resources to create enforcement mechanisms for all laws and to support all firms with legal compliance in the short term. Therefore, policies that use other frameworks for social upgrading are also important. They could give incentives to firms to comply with the law, facilitating law enforcement. Moreover, these frameworks could attract the leaders to go over what is required by law and bring even larger benefits to local development as compared with a reliance on law enforcement mechanisms alone. Therefore, public policy could help the governance of clusters and link small firms to global value chains that support transfer of knowledge along the chain to facilitate social upgrading through technical and financial resources (see chapter by Ipiranga, Faria and Amorim in this book). Policies could also spread the CSR concept and help firms to implement CSR initiatives to socially upgrade.

Clusters remains a good unit of analysis to think through policy solutions aimed at local development. Among the units of analysis of industrial organizations or economics (firms, sectors or chains), clusters are closest to encompassing the notions of territorial or geographical boundaries, which is fundamental when one analyzes social and environmental issues and local and/or regional development.

Many of the analyses on traditional upgrading can be applied to social upgrading. Externalities and initiatives for improving collective efficiency can also assist social upgrading within clusters, as they can provide:

- economies of scale for finding solutions (one solution that can be used by all);
- opportunities for collective joint action (e.g., common sewage treatment plants);
- development of specialized skills in the firms;
- potential for innovation in technology;
- spread of information and learning;
- potential for development of external services (consultancy, maintenance); and,
- scale of organization for social movements (environmental, labor) and law enforcement.

However, being in a cluster may also have some limitations for social upgrading, as clusters may lead to:

- complexities in finding solutions (scaling-up the problem because of the number of actors);
- higher costs/investments due to the larger scale;
- higher risk of negative impact on local development and disturbance of the cluster dynamics;
- political resistance and complicity (negative social capital) to bar change; and,
- greater opportunity for the devil's deal.

Those opportunities and limitations for cluster upgrading need to be investigated further in order to better understand the advantages of working with clusters for social upgrading. This book brings together a series of empirical studies that point to preliminary answers for many important questions that can shed new light on the concept of social upgrading. Three questions were selected to be answered (see conclusions, Chapter 8):

- How and why were clusters able to socially upgrade without losing competitiveness?
- What were the roles of external actors, such as governments, NGOs and clients in cluster upgrades?
- What were the key factors determining the effectiveness of the external interventions?

Thus, our main concern is how to move the research and policy agenda on clusters to address local sustainable development, or social upgrading, as a goal. This means strengthening the formal sector by facilitating small business incentives, registration and tax collection, respecting and enforcing labor and environmental regulations, and having more positive social impacts on local development. Many clusters have been successful at social upgrades, even within the existing regulatory burden. The research and practice community can better inform policymakers by understanding lessons from cases of effective cluster upgrades. This, in turn, can enable us to formulate new alternatives for development that go beyond the debates on regulatory reform, and to think on about how to reach sustainability for businesses, communities and the environment.

The next chapters of the book examines the discussions above empirically. They analyze cases of cluster social upgrading in Latin America and Asia and look at the factors that enable or prevent upgrading in the innovation capacity or in the labor, social and environmental standards.

In the next chapter, Anne Posthuma analyzes the opportunities for "high road" upgrading in a globalizing economy. She studied the insertion of SMEs of the Indonesian wood furniture sector in an increasingly competitive market driven specially by Chinese furniture firms. The Indonesian case offers an example of how globalization fostered favorable conditions for an industry sector to grow, internationalize and diversify, but many challenges appeared as new lower production costs enter in the global markets. The study provides the endogenous and exogenous factors that drive upgrading and sustainability in the furniture sector.

Puppim de Oliveira moves to the debates on how small firms in developing countries respond to the social and environmental standards when they are connected to global value chains. He studied how the exporters in the furniture sector in Brazil perceive environmentally-related demands from buyers abroad. Firms tend to have different responses according to the kind of product and exporting market, as well as the local environmental enforcement demands. International standards are becoming increasingly important proportionally to the percentage of the production exported.

Chapter 4 by Damiani explores the cases of social upgrading in agricultural clusters in China and Mexico. The tobacco cluster in the Yunnan Province in the People's Republic of China and organic coffee production in the State of Chiapas in Mexico were able to become more competitive generating economic development, while improving their social, environmental and labor standards. He identifies the main factors that helped to socially upgrade those clusters, such as the emergence of global standards and the institutional arrangements for governing upgradings.

The following chapter by Aya Okada analyzes how innovation is related to labor upgrading in knowledge-intensive clusters in developing countries. She examines the case of the improvements in labor capacity in SMEs in the Indian information technology (IT) industry and how a high quality labor force is part of their innovation strategies to upgrade. The chapter shows the diverse patterns of small firms' involvement in the global software industry, showing how SMEs can be dynamic organizations for innovating.

Mansueto Almeida examines how certain localities break the devil's deal to enforce regulations on clusters of small firms in developing countries. He looks at the important role of external actors to spark law enforcement on small firms, and at the same time support technological solutions to drive innovations and law compliance. The author examines the case of the jeans cluster in the dry municipality of Toritama in Brazil, and how the public attorneys pressed firms to comply with environmental laws on water resources on the one hand, and on the other hand the university provided the technological solution to water recycling and sewage treatment.

Finally, the chapter by Ipiranga, Faria and Amorim presents research on the empirical results of social mobilization technology as a tool of cluster governance to socially upgrade clusters of low technology, which generally have poor social, labor and environmental standards, as well as limited innovation capacity. It uses the experience of several years of the Nos Network in Northeastern Brazil to drive lessons on methods to mobilize clusters for catalyzing local development. Intangible aspects, such as social capital, cooperative practices and governance have shown to be fundamental to good cluster governance in order to generate economic and social improvements at the local level.

References

Altemburg, T. and Meyer-Stamer, J. (1999). How to Promote Clusters: Policy Experiences from Latin America. *World Development*, 27(9), 1693–1713.

Altemburg, T. and Eckhardt, U. (2006). *Productivity enhancement and equitable development: challenges for SME development*. Vienna, Austria: United Nations Industrial Development Organization (UNIDO).

Amorim, M. Alves (1998). *Clusters como estratégia de desenvolvimento industrial no Ceará.* Fortaleza, Brazil: Banco do Nordeste.

Anderson, Terry A. and Leal, Donald R. (1992). Free Market Versus Political Environmentalism. *Harvard Journal of Law and Public Policy*, 5(2), 297–310.

Asheim, B.T., Cooke, P. and Martin, R. (2006). The rise of the cluster concept in regional analysis and policy: a critical assessment. In B.T. Asheim, P. Cooke and R. Martin (eds), *Clusters and Regional Development: Critical Reflections and Explorations* (pp. 1–29). London and New York: Routledge.

Atzema, O. & E.J. Visser (2007). Innovation policy in regions with(out) clusters: A plea for a differentiated and combined network approach. In S. Hardy, L.B. Larsen & F. Freeland (eds), *Global Regions?* (pp. 14–17). Seaford: Regional Studies Association, Seaford.

Cassiolato, J.E., Lastres, H., Szapiro, M. (2000). Arranjos e Sistemas Produtivos Locais e Proposições de Políticas de Desenvolvimento Industrial e Tecnológico. Paper presented in the Seminar Local Clusters, Innovation Systems and Sustained Competitiveness, IE-BNDES, Rio de Janeiro.

Cassiolato, José Eduardo & Szapiro, Marina (2003). Uma caracterização de arranjos produtivos locais de micro e pequenas empresas. In Helena M.M. Lastres, José E. Cassiolato and Maria Lúcia Maciel (eds), *Pequena empresa: cooperação e desenvolvimento local*. Rio de Janeiro, Brazil: Relume Dumará Editora.

Cici, Carlo and Ranghieri, Federica (authors), Peinado-Vara, Estrella and de la Garza, Gabriela (eds) (2008). *Recommended Actions to Foster the Adoption of Corporate Social Responsibility (CSR) Practices in Small and Medium Enterprises (SMEs)*. Washington D.C., USA: InterAmerican Development Bank (IDB).

Cossentino, F., Pike F. and Sengenberger, W. (1996). Local and Regional response to Global Pressure: The Case of Italy and its Industrial Districts. Manuscript, International Institute for Labour Studies. Geneva.

Damiani, Octavio (2003). Effects on employment, wages, and labor standards of non-traditional export crops in Northeast Brazil. *Latin American Research Review*, 38(1), 83–112.

Galli, Rossana and Kucera, David (2004). Labor standards and informal employment in Latin America. *World Development*, 32(5), 809–28.

Gereffi, Gary, John Humphrey and Timothy Sturgeon (2003). The Governance of Global Value Chains, manuscript, November 4.

Gereffi G. (1999). International trade and industrial upgrading in the apparel commodity chain. *Journal of International Economics*, 48, 37–70.

Giuliani, Elisa, Pietrobelli, Carlo and Rabellotti, Roberta (2004). *Upgrading in Global Value Chains: Lessons from Latin American Clusters*. Quaderno n°72. Dipartimento di Scienze Economiche e Metodi Quantitativi. Università del Piemonte Orientale

Humphrey, J. and Schmitz, H. (2000). *Governance and Upgrading: Linking Industrial Cluster and Global Value Chain Research*. IDS Working Paper 120, Brighton, Institute of Development Studies, University of Sussex.

Humphrey, John, and Hubert Schmitz (2002). How does insertion in global value chains affect upgrading in industrial clusters? *Regional Studies*, 36(9), 1017–27.

Humphrey, John (2003). *Opportunities for SMEs in Developing Countries to Upgrade in a Global Economy.* Series on Upgrading in Small Enterprise Clusters and Global Value Chains, International Labour Office.

IBGE—Instituto Brasileiro de Geografia Estatística (1997). *Pesquisa da Economia Informal Urbana—ECINF 1997.* Rio de Janeiro: IBGE.

IBGE—Instituto Brasileiro de Geografia Estatística (2003). *Pesquisa da Economia Informal Urbana—ECINF 2003.* Rio de Janeiro: IBGE.

Kennedy, Lorraine (1999). Cooperating for survival: Tannery pollution and joint action in the Palar Valley (India). *World Development,* 27(9), 1673–91.

Markusen, Ann (1996a). The Interaction between Regional and Industrial Policies: Evidence from Four Countries (Korea, Brazil, Japan, and the United States). Proceedings, Annual World Bank Conference on Development Economics, 1994, pp.279–98.

Markusen, Ann (1996b). Sticky places in slippery space: A typology of industrial districts. *Economic Geography,* 72, 293–313.

Marshall, A. (1890). *Principles of Economics.* Londres: MacMillan and Co.

Martin, Ron and Peter Sunley (2003). Deconstructing clusters: chaotic concept or policy panacea? *Journal of Economic Geography,* 3, 5–35.

Mead, Donald C. and Carl Liedholm (1998). The dynamics of micro and small enterprises in developing countries. *World Development,* 26(1), 61–74.

Meyer-Stamer, J. (1998). Path Dependence in Regional Development: Persistence and Change in Three Industrial Clusters in Santa Catarina, Brazil. *World Development,* 26(8), 1495–1511.

Nadvi, K. and Waltring, F. (2004). Making sense of global standards. In H. Schmitz. (ed.), *Local Enterprises in the Global Economy: Issues of Governance and Upgrading.* Cheltenham, UK: Edward Elgar.

Nooteboom, B. (2006). Innovation, learning and cluster dynamics. In: R. Martin, B. Asheim and P. Cooke (eds), *Clusters and Regional Development; Critical Reflections and Exploration.* London: Routledge, pp. 137–63.

Okada, Aya. (2004). Skills Development and Inter-firm Learning Linkages under Globalization: Lessons from the Indian Automobile Industry. *World Development,* 32 (7), 1265–88.

Piore, M.J., Sabel, C.F. (1984). *The Second Industrial Divide: Possibilities for Prosperity.* New York: Basic Books.

Porter, Michael E. (1998). Clusters and the new economics of competition. *Harvard Business Review,* 76(6), 77–90.

Porter, Michael E. (1996). Competitive advantage, agglomeration economies, and regional policy. *International Regional Science Review,* 19 (1&2), 85–94.

Posthuma, Anne Caroline (2004). Industrial Renewal and Inter-firm Relations in the Supply Chain of the Brazilian Automotive Industry. *Series on Upgrading in Small Enterprise Clusters and Global Value Chains,* ILO Working Paper 46.

Powell, Watler W. (1990) Neither market nor hierarchy: network forms of organization. *Research in Organizational Behavior,* 12, 295–336.

Puppim de Oliveira, Jose A. (2002). Implementing environmental policies in developing countries through decentralization: The case of protected areas in Bahia, Brazil. *World Development,* 30(10), 1713–36.

Puppim de Oliveira, Jose A. (2008). *Implementation of Environmental Policies in Developing Countries: A Case of Protected Areas and Tourism in Brazil.* Albany, NY, USA: State University of New York Press (SUNY).

Pyke, F., Becattini, G. & Sengenberger, W. (1990). *Industrial Districts and Inter-Firm Co-Operation In Italy.* Geneva: International Institute for Labour Studies.

Rabellotti, R. (2001). *The Effect of Globalisation on Industrial Districts in Italy: The Case of Brenta,* Global and Local Governance Programme Project Report, Brighton, Institute of Development Studies.

Rabellotti, R. and Schmitz, H. (1999). The internal heterogeneity of industrial districts in Italy, Brazil and Mexico. *Regional Studies,* 33, 97–108.

Samstad, James G. and Seth Pipkin (2005). Bringing the firm back in: Local decision making and human capital development in Mexico's Maquiladora Sector. *World Development,* 33(5), 805–22.

Schmitz, H. (1995). Small shoemakers and fordist giants: Tale of a supercluster. *World Development,* 23(1), 9–28.

Schmitz. H. (ed.) (2004). *Local Enterprises in the Global Economy: Issues of Governance and Upgrading.* Cheltenham, UK: Edward Elgar.

Schmitz, H. (2005). *Value Chain Analysis for Policy Makers and Practitioners.* Geneva, Switzerland: International Labour Office (ILO).

Schmitz, Hubert and Bernard Musyck (1994). Industrial districts in Europe—Policy lessons for developing countries? *World Development,* 22(6), 889–910.

Schmitz, H. and Nadvi, K. (1999). Clustering and industrialization: Introduction. *World Development,* 27(9), 1503–14.

Tendler, J. (2002). Small Firms, the Informal Sector, and the Devil's Deal. IDS Bulletin [Institute of Development Studies], Vol.33, n.3, July.

Tendler, J. and Amorim, M. A. (1996). Small firms and their helpers: lessons and demand. *World Development,* 24(3), 407–26.

Tewari, Meenu, and Pillai, Poonam (2004). Global Standards and Environmental Compliance in the Indian Leather Industry. *Oxford Development Studies,* 33(2), 245–67.

Uhlig, Christian (1992). Environmental Protection and Economic Policy Decisions in Developing Countries. *Intereconomics,* March/April.

Visser, E.J. & de Langen, P. W. (2006). The importance and quality of governance in the Chilean wine industry. *GeoJournal,* 6, 177–97.

Vives, A. (2006). Social and environmental responsibility in small and medium enterprises in Latin America. *Journal of Corporate Citizenship,* Issue 21, 39–50.

Chapter 2

Seeking the High Road to Jepara: Challenges for Economic and Social Upgrading in Indonesian Wood Furniture Clusters

Anne Caroline Posthuma[1]

2.1 Introduction

Global production by transnational corporations (TNCs) has become a major driver of economic growth and trade expansion worldwide. TNCs are estimated to account for around two-thirds of total world trade (UNCTAD 2001) and the world value of exports has doubled between 2000 and 2006 (UN 2007:37), reflecting this deepening of global economic integration.

Developing countries have been drawn more heavily into world trade as a result, with output growth rising at its highest rates in decades (ibid.:1). Of concern to governments and companies in developing countries is whether this rising output and export participation will translate into new and sustainable opportunities for local firms to grow, become more competitive and create new jobs. From a development perspective, trade and foreign direct investment (FDI) bring valuable opportunities for economic upgrading of local firms via transfer of technology and new technical know-how, exposure to new management and production techniques, quality control procedures and access to new markets. With the economic upgrading of local firms arise concerns about whether new employment created will be in low-skilled, low-wage jobs; or if they can be considered quality job opportunities where employers provide a formal employment contract to workers, pay a living wage on a regular basis, respect the right to organize in the workplace, without involving excessive overtime in accordance with national labor legislation, including provision of social security and respecting occupational safety and health standards.

Case studies examining the impact of trade and Foreign Direct Investment (FDI) on economic upgrading and/or social upgrading in specific sectors have shown that outcomes can be mixed. While rising output and economic upgrading has been attained by some local small and medium enterprises (SMEs), this is not always accompanied by upgrading of social and labor conditions for workers and improved environmental management. There is rising awareness within

---

1   The author is Senior Research Officer at the International Institute for Labour Studies of the International Labour Office. The views expressed in this chapter do not necessarily reflect those of the ILO or of its constituents.

the development community and institutions that equitable and sustainable economic growth should also include respect for labor standards and should avoid environmental degradation[2].

At first glance, the growth of the Indonesian wood furniture industry in recent decades suggests that this local craft-based sector may provide an exemplary case of how globalization has fostered favorable conditions for a developing country industry to internationalize, access new markets, raise output, diversify, upgrade their production techniques and create new jobs. Indeed, local furniture production was revived in the 1970s and rapidly penetrated global markets in less than 30 years. This lead to an intense process of local industrial expansion and strengthening of inter-firm relationships within SME clusters that have specialized production, carving and marketing skills. However, a deeper examination shows that fragilities in the structure of production and poor labor practices which already existed in this local industry might be intensified with the impact of global production. Furthermore, a survey with global buyers showed their high rating of criteria such as low product price, high product quality, quick and reliable delivery times and good client relations, which forms the basis of increasing competition between Indonesian SME producers of furniture with other furniture-producing Asian countries. The restructuring decisions of these clusters will determine whether price-based, labor-intensive strategies will prevail, or whether it will be possible to upgrade into higher quality, skill- and design-intensive furniture markets. Furthermore, the rapid depletion of Indonesian forests of teak and mahogany requires that sustainable approaches to natural resource management be adopted by clustered firms, including new production practices that use scarce precious hardwoods efficiently, respecting controls on illegal logging, using certified sustainably-grown timber and combining new materials with traditional designs.

Both endogenous and exogenous factors are involved in creating the pressures and opportunities for Indonesian furniture clusters to restructure. In order to examine those issues empirically, this chapter analyzes the case of the furniture cluster in the region around the town of Jepara, Central Java Province. This case study raises development questions regarding whether participation of developing country clusters in global production, especially clusters supplying to value chains with a quasi-hierarchical governance structure, can lead to a virtuous cycle where expanded production and new export opportunities can be harnessed to upgrade the industrial competitiveness of firms while also supporting environmentally

---

2   For example, the International Finance Corporation (IFC) has recently included labor standards provisions in their lending practices, and the Equator Banks have now included Performance Standard 2 in their lending programs, meaning that now 70% of emerging market finance now has a labor standards requirement. In the same vein, the Millennium Development Goals now include, as part of Target One to eradicate extreme poverty and hunger, the achievement of "…full and productive employment and decent work for all, including women and young people".

sustainable practices and improving the quality of employment and working conditions for workers in this sector.

2.2 Strengthening SME Clusters in the Context of Global Competition

SMEs are an important source of job creation in developing countries and frequently receive special policy attention. Indonesia is no exception, as small enterprises play an important social and economic role, employing over 60% of the Indonesian workforce (Indonesian Central Bureau of Statistics, 2002). As a result, the ability of SMEs to upgrade by growing, generating profits for reinvestment and adopting new production practices all bear heavily upon the scale and quality of jobs created by these firms.

Nevertheless, sustained development of SMEs is a challenge because their mortality rate is high during the first five years of existence, as many firms cannot withstand competition and fail to grow. Second, SMEs are often associated with low levels of investment, weak technology transfer and mastery, low productivity and inefficient work practices. Third, studies show that SMEs may operate within a poor working environment, may not comply with occupational safety and health regulations and often rely upon informal employment practices. Fourth, many SMEs are not formally registered firms, which restricts their eligibility to apply for credit and other financial and business services.

There is a need for government policies that boost SMEs in ways that unleash their potential for dynamic job creation and improve productivity, competitiveness and labor conditions. Yet, industrial policies tend to be directed toward larger companies, whereas policies for SMEs are often placed within the framework of social policies and poverty-alleviation measures that involve subsidies and micro-credit (Tendler, 2002).

A core problem for many SMEs is their isolation. Bolstering the embedding of SMEs in productive relationships can generate a number of externalities, such as reduced transaction costs, improved bargaining power with suppliers of raw materials as well as with buyers and unleashing collective efficiencies. The benefits to be gained through more intense horizontal inter-firm linkages have been long recognized, described in the early twentieth century by Alfred Marshall and more recently in the substantial body of literature on clusters and industrial districts that points toward superior competitiveness of firms that are territorially embedded. The literature on clusters and industrial districts presents an alternative production model that places SMEs as protagonists in attaining high growth rates, displaying innovative behavior and competing successfully in domestic and international markets. The interest in clustering has risen, especially among policymakers, as the evidence suggests that clustered firms are more resilient in periods of economic fluctuation than large firms and their non-clustered counterparts (Schmitz, 1989; Schmitz and Musyck, 1994). This perspective of clustered firms has opened new scope for SME policies—rather than considering SMEs as a marginal category, composed of mainly survivalist production units or representing a transition stage on the path to "grow up" into large firms, the evidence raised in many case studies shows that SMEs can become an integral part of regional patterns of industrial development, and can play a potential role in incipient industrialization (Schmitz, 1989).

It is useful to identify some primary characteristics that distinguish clustered firms from mere agglomerations of business activities in the same territory. Indeed, not all agglomerations of small firms produce growth, stable employment and technical innovation, showing that such outcomes are not automatic and cannot be attributed solely to market factors. We highlight three features of clusters here. First of all, clustered SMEs demonstrate productive features that lead to economic growth and high specialization of production in small batches (versus mass production). Highly skilled workers are an essential feature, and are related to the ability to establish a division of labor among firms that enable them to capture economies of scale external to the individual firm but internal to the cluster, while innovative activities and technical and design skills give rise to economies of scope in product lines (Brusco, 1982; Piore and Sabel, 1984, Schmitz, 1989). Second, more subtle features strengthen and support competitiveness of small firms, as they are embedded within formal and informal economic, social, cultural and political institutions and relations of the territory or area which define the community and its institutions of local governance within the cluster (Pyke, Beccattini and Sengenberger, 1990; Locke, 1995:3; Criscuolo, 2005; Bellandi and Caloffi, 2007:4). "In the district, unlike in other environments, community and firms tend to merge (leading to a) thickening of industrial and social interdependencies" (Becattini, 1990:38). Finally, the coexistence of inter-firm cooperation together with competition generates a creative tension where entrepreneurs are willing to engage in collective joint action, leading to collective efficiencies arising from their interdependence, rather than rent-seeking behavior, while also following market signals to upgrade their products and win new clients.

While the early clustering literature tended to build an "ideal type" based upon the Italian experience, subsequent studies pointed toward the heterogeneity among clusters (Amin and Robins, 1990) and recognized the difficulty of replicating this model of development in other industrial settings (Piore, 1990: 225–7). Indeed, the recognition of regional, sectoral and institutional specificities within clusters in developing countries has facilitated cross-country comparisons, as well as sharing of lessons learned and policy experiences.

Finally, it is important to consider the implications for decent work within clusters in developing countries and the challenges for social and labor upgrading, especially in the context of globalization. While small-firm production systems generate jobs, there are challenges associated with the quality of some of these jobs, where labor standards may not always be abided by and where informal work is used frequently. The implications of job creation in clusters operating under labor surplus conditions, such as those found in developing countries, were already raised in early studies of industrial districts and clusters in the industrialized countries (Locke, 1995; Sengenberger and Pyke, 1992; Schmitz, 1989), cautioning that this could possibly lead to a "low road" path of reducing wage costs in order to maintain competitiveness. Meanwhile, some authors emphasize that SME clusters should adopt a "high road" strategy, characterized by product quality and sustained innovation. Still, other authors suggest that a middle path may emerge where competition based upon production innovation, quality and price coexists with labor surplus conditions (Nadvi, 1994:193). The case study examined in this chapter suggests that when global buyers generate price-based competition between their pools of suppliers in different countries, this creates pressures that make it difficult for firms in developing countries to adopt a "high road" competitive strategy.

As the process of globalization unfolds, many clustered firms in developing countries have started supplying to global buyers, in addition to their traditional local and national markets. As a result, attention by researchers and policymakers has been directed toward the role of global production networks in helping SMEs in developing countries to upgrade, acquire new technologies, learn new techniques and quality control methods and access new export markets. A number of recent studies have examined the relationship between clusters of SMEs in developing countries with buyers through global value chains. These studies have examined the opportunities and risks for SMEs in developing countries to participate in the global economy in a variety of sectors and have considered policy options to prevent negative outcomes (World Development, September 1999 and IDS Bulletin, July 2001). Of particular interest is a typology of four different forms of governance of value chains by global buyers and the implications raised regarding the scope for supplier upgrading in each (Humphrey and Schmitz, 2001) that will be presented later in Section IV.

There is a need to identify policies and strategies to help SME clusters reap gains from participating in global production networks. In this context, it remains a core question whether "high road" upgrading is possible, where competitiveness can also involve good conditions of work, or decent work (ILO, 1999). The following section discusses research findings on SME furniture clusters in Central Java, and considers how participation in global production has impacted upon their industry, the jobs created and the sustainability of their raw material inputs of high-quality timber.

2.3 Origins of Jepara's Woodworking Prowess

The town of Jepara is linked with a craft tradition in Indonesia, where even local legends recount how past kings magically granted Jepara's citizens their great woodworking skills. Historical records confirm that a wood-carving industry has long-existed in the Jepara region, and that the skills, aptitudes, values and tools necessary for a craft industry have survived over time and through periods of political and economic upheaval.

By the sixteenth century, Jepara was already a thriving commercial center. Yet, the town slipped into obscurity in the seventeenth century, after the Dutch burnt the town and moved the administrative center to Semarang. After a long dormant period, the local industry was resuscitated in the 1970s through a revival of domestic demand and renewed appreciation for traditional-style furniture. With this growth, Jepara re-emerged as the hub of a burgeoning industry, encompassing approximately 3,000 firms, including 100 large and medium enterprises scattered across 80 villages (Sandee et al., 2002: 15). Employment in the woodworking sector was estimated at 44,000 workers by the late nineties, but this total employment could easily double, if one considers those informal workers, who work as casual laborers, often without any contract or social protection, and are paid on a piece-rate basis (Schiller and Martin-Schiller, 1997:21). As such, informal work is an integral feature of the cluster; a key question of concern to this chapter is whether participation of these furniture clusters in global production is creating the conditions that can improve the quality in which people work. This issue will be taken up later in this chapter.

From a modest initial attempt at exporting wood furniture from Central Java Province in 1986, the value of exports had reportedly reached US$169 million in 1998 (ibid.)[3]. The 32 furniture clusters in Central Java Province were responsible for 21.6% of the Province's total exports in 2001 (although this represented a decline from 27% in 2000), as compared with 13.2% for garments and 13% for textiles. The importance of the furniture industry as a source of economic activity, export earnings, job creation and income generation is greatest at the provincial level. When one considers the national aggregate of economic activity, the furniture industry represented only 1.87% of total manufacturing outputs in 2002, and added around 2.7% to the total value of Indonesian exports (CEMSED, 2003:3 and 4).

The wood furniture sector in Indonesia offers a compelling case of the opportunities and challenges raised by globalization for SME producers in developing countries. While a large part of Central Java's export success can be attributed to a number of favorable circumstances, both internal and external to the wood furniture clusters, the sustainability of this export success has been put under pressure by price-based competition from other countries in the region, especially China and Vietnam. The main exogenous and endogenous factors involved are examined below.

Two main exogenous factors at play can be highlighted. First of all, the rise in global demand for hand-crafted furniture was the initial factor that encouraged

---

3   It is likely that government statistics underestimate the value of Jepara exports, as many businesses under-report the value of exports to reduce their tax burden. Furthermore, government statistics also omit indirect exports which exclude substantial amounts of furniture partly processed in Jepara and completed elsewhere, which could add as much as 75% to the official export total. Similarly, government figures underestimate the furniture industry's labor force, as data is not collected from unregistered companies or for workers hired on a piecework basis.

the rise of Jepara furniture exports. However, subsequent changes in the sourcing practices of the major global buyers in the wood furniture sector, particularly the search for cheaper products in lower labor-cost countries in the Asian region, has been a key determinant of whether market opportunities are opening or closing, and where new sourcing contracts will be signed. For example, the share of exports of secondary processed wood products from developing countries overall to the OECD dropped from 17% in 1996 to 15.4% in 2000, whereas China increased its share from 22% in 1996 to 35.7% in 2000 (Tissari, 2002:10). Few countries can compete with China on wage costs. Yet, when one compares labor productivity in the wood furniture sector in 2000 between Indonesia ($28,000 per employee, per year) and Italy ($89,000 per employee, per year) (Tissari, 2002:23), this suggests that Indonesian wood furniture production might fall between two stools — neither competitive as a low labor-cost zone nor as a high labor-productivity zone.

Second, policy reforms by the Indonesian government that opened the domestic market to international trade have been important. In 1986, the Indonesian central government lifted regulations upon wood furniture exports, further boosting the production and exports of this sector. The government also invested in infrastructure improvements, including the upgrading of regional harbors such Semarang and some local roads; all of which greatly reduced the transport distance and time to get furniture from Jepara to vessels where they would be transported to markets overseas[4]. Export procedures were eased, so that by 1996, approval was only required from the shipping agent and surveyor who conducted the final inspection prior to checking and sealing the shipment (Schiller and Martin-Schiller, 1997:9). The depreciation of the Rupiah to the dollar due to the Asian financial crisis of the late 1990s provided a temporary boost to exporters, especially as they had a low import coefficient, because most raw material inputs were available locally. At the same time, this depreciation raised local costs and led to a decline in domestic sales of wood furniture. The government sponsored several regional, national and international furniture exhibitions that boosted international awareness of Jepara furniture and brought many new foreign entrepreneurs to the country. Finally, the Jepara government placed a tax on the shipment of unfinished timber from Jepara, and the central government banned the export of logs in 1987, thereby creating effective protection of the domestic timber industry for domestic producers[5]. However, this export ban was lifted in the late nineties, in response to IMF policy advice, which further fueled illegal logging of national forests and generated higher costs and supply shortages for local producers to access timber.

---

4   Although some infrastructure improvements made during the 1980s were not maintained during the 1990s, including a deterioration of Semarang harbor (Sandee et al., text with no date).

5   This effective protection provided a crucial boost to the local furniture industry, as timber accounted for nearly half of Jepara production costs (Schiller and Martin-Schiller, 1997:9).

Several main endogenous factors have been responsible for both helping and hindering the competitiveness of this industry. First of all, as exports rose rapidly, some producers in Jepara were unable to meet their production targets and delivery schedules. Many companies reportedly lost their business credibility with clients as a result. Second, as the town of Jepara grew and become crowded, this served as the springboard for the creation of other clusters in Central Java Province, as some businesses relocated to other nearby towns. Third, a rivalry between local producers and some foreign exporters led a few foreign-owned firms to relocate away from the town.

Foreigners have played an important role in the development of this local industry. On the one hand, they have tended to enter and dominate the higher value-added activities such as exports, which have led to resentment among some local Indonesian producers and traders. On the other hand, foreigner traders have provided access to important markets overseas, have helped upgrade the quality of local furniture production and have facilitated access to working capital by extending partial advance payment for export transactions, which could help local producers to cover part of their initial costs of timber and wages. Foreigners also introduced new designs and helped to diversify the overall range of designs available, as local firms imitated and sought to differentiate themselves on the basis of timber quality, finishing techniques and design. As a result, Jepara developed a reputation for being flexible in the production of "antique" reproduction furniture. This also has involved violation of copyrights and exclusive designs, as Jepara producers put designs on the Internet, making them available to other producers and buyers (Sandee, et al., no date on text). Finally, the growth of the local industry has brought about a much denser network of clustering activities among furniture producers. This has led to the emergence of a few large firms, which have taken the role of lead firms over a chain of local suppliers. Firms within clusters have been able to specialize by actively outsourcing parts to smaller firms. A survey of clusters in Central Java Province revealed dense inter-firm relations within clusters, as well as intense buying between clusters in the Province (CEMSED, 2002)[6]. This process has involved an intensification of flexible labor practices, where existing informal forms of labor use have been intensified as SMEs have grown, specialized and used outsourcing more intensively (ibid.).

A general conclusion of this analysis was that Jepara was in a transition process at the turn of this century and as a result of greater participation in global markets, moving toward the rising participation of Indonesian businesses in both production and marketing. In this process, some labor was shed, and researchers documented

---

6   Previous studies have used Jepara as their case study for the Indonesian furniture industry. However, this cluster survey (CEMSED, 2002) included three issues in the analysis, which had been absent in previous studies, namely: (a) the dynamic linkages between clusters in all of Central Java; (b) the use of value chain analysis to examine how clusters are related to global buyers; and (c) much greater and comprehensive analysis of employment and working conditions than in previous studies.

the impacts of increased cost competition that raised negative consequences for labor standards and working conditions (Sandee, Sulandjari and Rupidara, 2002).

The next section of this chapter adopts a value chain analysis in order to better illustrate the range of challenges for local industries in developing countries that participate in global markets. Attention is directed toward three aspects of the production chain in the wood furniture industry. First of all, the pressure from global markets in search of ever-cheaper mass-produced goods is considered and whether niche markets may provide an important strategy for diversifying production. Second, labor practices in these firms are examined and the question is raised whether increased competitive pressures will lead to a higher degree of unregistered firms (which was found to be at 83%) and use of informal labor (which was found to be at 67%), or whether a "high road" option of improved competitiveness, with good business and labor practices, might be a viable option. Finally, the upstream constraints surrounding the restricted supply and rising cost of teak and mahogany to this industry are considered and whether this will create a fatal bottleneck to the sustainability of this industry, in the absence of policy interventions to sustainably manage the limited remaining forest resources in the country.

2.4 Main Findings and Challenges

The following section draws upon background research conducted for a program of activities[7] which were completed in 2005. The discussion will first address some issues related to downstream linkages in the value chain (i.e. with global buyers) and then upstream linkages related to problematic access to raw material inputs and national forestry management. We then turn attention to the informality of both individual firms in the Central Java clusters and of their hiring practices. The final section will highlight some initial considerations for policy that arise from these preliminary research findings.

The Indonesian wood furniture industry is a particularly interesting case to examine in terms of SME upgrading possibilities in a globalizing world. First of all, it is an industry that has experienced a substantial rate of increase in global trade since the 1990s (Kaplinsky, Morris and Readman, 2002). Second, it is a buyer-driven chain, which enables us to examine the role of global buyers in creating and limiting opportunities for increased access to export markets and functional upgrading by local SMEs. Third, wood furniture is a resource-intensive industry, which raises important implications about upstream bottlenecks and environmental sustainability of such industries if the local resource base is not

7   The preliminary research findings were gathered and the cluster survey was conducted as part of project activities conducted by the Job Creation and Enterprise Development Department of the ILO between 2002 and 2005 and completed since that time.

managed wisely. Finally, wood furniture is an important industry for Indonesia and a core industry in Central Java Province, accounting for substantial job creation, income generation, production and export earnings. By 2000, furniture production contributed between 1 and 1.87% to total Indonesian manufacturing output, and around 2.7% to the total value of Indonesian exports. In Central Java Province, wood furniture was the largest contributor to provincial exports, accounting for 27.16% of total exports in 2000 and although this dropped to 21.57% in 2001, this sector still accounted for the highest share of total exports (versus garments with 13.28% and textiles with 12.80%) (CEMSED, 2002:3–4).

2.4.1 Downstream Linkages in the Wood Furniture Value Chain

Trade data from UNCTAD on exports of wood products from selected Asian countries[8] reveals that substantial growth was attained during the 1990–95 period, but growth rates of exports declined substantially in the same set of countries during the 1995–99 period (Readman, 2002:9). It is worth noting that the growth rate slowed even in China. Hence, this slowdown in exports is not a problem confined to wood furniture producers in Central Java Province, but more generalized to other furniture-producing countries in the region. Impacts from the Asian financial crisis may not be a relevant explanation in this case, as exported products were generally found to have benefited from exchange rate devaluation resulting from the crisis. Instead, one might hypothesize that this slowdown reflects more structural factors in the wood products industry, possible involving world market factors, changes in buyer preferences, changes in sourcing decisions, or saturation of markets.

It is useful to bear in mind the extensive structure of the wood furniture value chain. A simplified representation of the value chain in the wood furniture industry is given in Figure 2.1.

Some clusters of wood furniture producers examined in Central Java Province include exporting firms. This trend to include marketing and export agents in the clusters has been encouraged by two local factors. First, since 1990, many foreign businessmen started operating in the furniture cluster in Jepara and some established a partnership with local entrepreneurs based upon a division of labor where the local partner would deal with production activities and the foreign partner would deal with marketing. Second, since 1997, the Indonesian government has opened the furniture subsector for foreign investment (CEMSED, 2002:8–9). However, this arrangement has not been entirely satisfactory in some cases and has contributed to a nationalist sentiment toward business practices. This sentiment is understandable when one gets an idea of the share of value-added obtained by wholesale and retail trade, as seen in Figure 2.2. The large share of revenues to be gained by traders also suggests that this is an important area for policy intervention to boost trading skills and capacity at the local and national

---

8   These countries are China, Indonesia, Malaysia, the Philippines and Thailand (UNCTAD, 2000 and 2001).

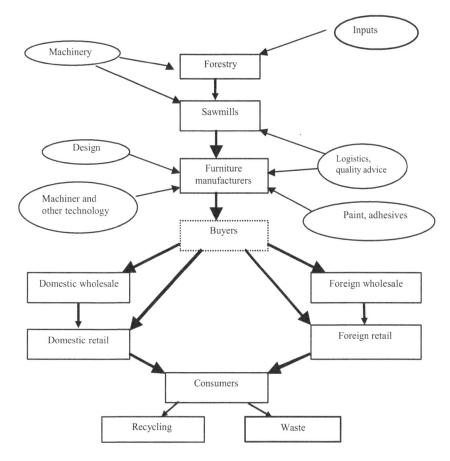

Figure 2.1   Wood furniture value chain

Source: Readman, 2002: 16, adapted from Kaplinsky, Morris and Readman, 2002.

level in order to increase value-added and revenues being captured by local firms, that could be reinvested in upgrading of production capacity, techniques as well as upgrading of employment and working conditions in these local companies.

Hence, the issue arises of who controls the value chain, and what types of opportunities arise for SMEs and their scope for upgrading and moving into higher value-added activities. A particularly useful approach for analyzing the relationship between different forms of buying relationships and the scope for cluster upgrading has been developed as a result of a multi-country study that analyzed local and global governance issues, based at the Institute of Development Studies. Four different types of linkages between SME clusters and global buyers have been identified by Humphrey and Schmitz (2001), as shown in the Table 1.1 in Chapter 1 of this publication.

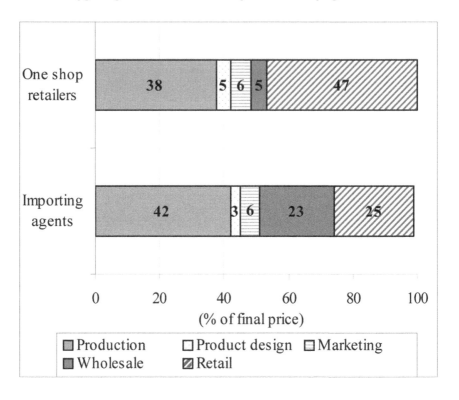

Figure 2.2   Distribution of value-added in the wood furniture value chain —the perspective of global buyers

Source: Readman, 2002:23.

Using the typology developed in the Table 1.1, most developing countries clusters are based in quasi-hierarchical relationships. Although such relationships are costly, require asset-specific investments in relationships with particular suppliers and also increase the rigidity of supply chains by raising the costs of switching suppliers, this type of structure nevertheless provides the global buyer with control over specification of product design. Many buyers are concerned to control this element, in order to avoid potential losses arising from a failure to meet commitments (for example, delivering the right product on time) or a failure to ensure that the product conforms to the necessary standards. Three main reasons why quasi-hierarchical relationships are a particular feature of the insertion of developing country firms into the contemporary global economy include (1) product differentiation and innovation are becoming increasingly important sources of competitive advantage, (2) final product markets in developed countries are characterized by an increasing emphasis on safety, labor and environmental standards, which requires greater monitoring and supervision of production processes and (3) in some sectors, there is a degree of task complexity and/or time pressure that requires coordination of tasks across firms (Humphrey, 2003).

It should be noted that a potential contradiction arises between the prevalent quasi-hierarchical sourcing behavior and the goal to upgrade Indonesian wood furniture clusters. This will be discussed further in the final section on policy implications.

2.4.2 Upstream Linkages—Raw Materials Utilization

Wood furniture is a resource-intensive industry, as noted earlier, drawing heavily on local wood resources in Indonesian forests. The clusters in Central Java purchase their wood from three main sources: from Perum Perhutani (the State-Owned Forest Enterprise); from traders; and from private forests. As seen in Table 2.1 below, the private traders were the key source of wood for the majority of firms in the clusters examined in the Center for Micro & Small Enterprise Dynamics (CEMSED) survey, both in terms of quantity as well as prevalence across the number of clusters interviewed. It is important to note the predominant use of the precious woods, teak and mahogany.

The Indonesian wood furniture industry emerged from its rich natural resource base in precious woods, and skilled artisans emerged from the need to work this wood. An interesting developmental implication arises in this regard.

What is the future developmental perspective of an industry that is dependent on its natural resource base, when that source of natural resources is under threat and being depleted at a rapid rate? As seen from the experience of the East Asian NICs (Newly Industrialized Countries), significant levels of local development and rising export performance, as well as the increased skill of the domestic labor force, is not necessarily tied to specific locally-available resources. However, such industries did emerge and grow under strong *dirigiste* government policies, aimed at supporting the development of indigenous industries and were helped by complementary policy support such as import protection, sources of finance including lines of credit, strategic licensing and transfer of foreign technology. It appears that such conditions would not apply to the Indonesian wood furniture sector under current policies favoring market liberalization; which leads us to question the sustainability of such a resource-intensive industry where its resource base is under threat, if no policy measures to support the transition to new wood types (which involves issues of supply, and possible changes in equipment, materials, skills and designs) are in place.

On the one hand, high local-content industries such as wood furniture production survived better than other firms during the Asian Crisis, as they did not suffer problems related to supply constraints and unfavorable exchange rates under the crisis (CEMSED, 2000:4). On the other hand, mounting evidence shows that Indonesian forests are being harvested at unsustainable rates (Environmental Investigation Agency, 2003). Given that these firms rely upon using precious woods such as teak and mahogany, it would be difficult to switch to imported wood

Table 2.1   Source of wood and wood type used by Central Java wood furniture clusters in 2002

Source of wood	Origin		
Perhutani	Teak Boyolali Blora Kendal Klaten Pemalang Rembang Sragen	Mahogany Magelang Kediri, Jawa Timur Jawa Barat	Others Batang Cilacap Jepara Tegal Wonosari, Yogya Jawa Timur
Traders	Boyolali Blora Jepara Kendal Klaten Pemalang Purwodadi Purworejo Rembang Salatiga Solo Sragen Subah, Batang Temanggung Wonosari, Yogya	Boyolali Gunung Kidul Jepara Pemalang Purwodadi Solo Sragen Tegal Wonogiri Jawa Timur Jawa Barat Yogyakarta	
Private forest	Purworejo Sragen Yogyakarta	Klaten Sragen Yogyakarta	

Source: Cluster survey, CEMSED (2002)

as a solution, because few other countries have indigenous supplies of mahogany and teak (and that which is available on the world market is more expensive than prices normally paid for these domestic woods).

One strategic response to this situation for companies and policymakers situation might be to concentrate upon developing niche markets for high-quality wood furniture. Such a strategy would involve some clear policy interventions, including:

- training of the labor force, both in factories and in the vocational training schools, in more sophisticated woodcarving skills suitable to more specialized and discerning markets, where heavy carved wood furniture is being replaced by more sleek designs;
- design training, perhaps developing a design school in the region, that would introduce new designs that reflect contemporary trends toward

lighter and more delicate models, including the mixing of fine woods with other materials such as metal to create new effects;
- technical support for improving marketing skills and ability to access new markets;
- business development services (BDS) provided to manage technological change, adapt to free trade conditions, and encourage local entrepreneurship;
- financial assistance, to help companies invest in new production equipment, to conform to more stringent quality standards and to just-in-time delivery requirements; and
- consultancy services (in commercial law, marketing information and in production technology such as kiln drying to prevent cracking of wood that has not been properly dried beforehand, use of new machinery, changing to new designs, etc.) (CEMSED Survey, 2002).

In terms of relevant interventions at the firm and cluster level, the preliminary research findings indicated a very high rate of waste and inefficient use of wood in the manufacturing process. A series of potential interventions could be developed in order to assist these firms to improve their work methods, streamline the production line and introduce more efficient resource management, which would contribute not only to increased productivity and reduced costs, but also contribute toward improvement of occupational safety and health in these workshops which often operate with very simple production procedures and machinery.

2.4.3 The High Rate of Informality among Firms and Hiring Practices

The Center of Micro and Small Enterprise Dynamics (CEMSED) survey of clusters in Central Java revealed that the majority of firms in these clusters were unregistered and the majority of their workers were unregistered. As can be seen from Table 2.2, 83% of firms interviewed in these clusters operated in the informal economy. Furthermore, their hiring practices relied upon the use of precarious labor, as more than two-thirds of the workers in the labor force had no formal employment contract and therefore no recourse to unemployment or health benefits, or any other form of social protection via the employment relationship. The research data also indicated poor and frequently unsafe working conditions in the clustered firms that were visited. These findings suggest that, in order to remain competitive and keep their operating costs lower, these firms avoid taxation by remaining unregistered. This has a negative impact upon their capacity to upgrade either economically or socially, as they cannot avail of existing support services and lines of financing or credit, which acts in turn as a constraint on the ability of the firm to invest, improve its production practices and human resources policies and to grow (Tendler, 2002). On a similar note, as regards environmentally sustainable business practices, the preliminary research findings did not indicate that wood furniture buyers were interested in encouraging the use of certified sustainably harvested timber in their

Table 2.2   Distribution of number of enterprise and number of workers in furniture clusters in Central Java, 2000

No	Region	Enterprise				Number of workers				Markets Served D=domestic E=export
		Informal	Formal	Total	%	Informal	Formal	Total	%	
1	Klaten	1,312	44	1,356	16.57%	3,406	300	3,706	9.56%	D, E
2	Grobogan	1,312	74	1,386	16.94%	3,406	482	3,888	10.03%	D, E
3	Jepara	956	70	1,026	12.54%	10,216	1,352	11,568	29.84%	D, E
4	Blora	235	173	408	4.99%	427	941	1,368	3.53%	D, E
5	Boyolali	235	16	251	3.07%	628	305	933	2.41%	D, E
6	Semarang	215	20	235	2.87%	315	775	1,090	2.81%	D, E
7	Sukoharjo	69	33	102	1.25%	235	579	814	2.10%	D, E
8	Kendal	10	80	90	1.10%	30	1,015	1,045	2.70%	D, E
9	Semarang	52	19	71	0.87%	166	340	506	1.31%	D, E
10	Purworejo	25	17	42	0.51%	25	328	353	0.91%	D, E
11	Salatiga	12	13	25	0.31%	55	54	109	0.28%	D, E
12	Surakarta	---	11	11	0.13%	---	93	93	0.24%	D
13	Sragen	1,276	60	1,336	16.33%	3,584	526	4,110	10.60%	D
14	Tegal	754	280	1,034	12.64%	2,854	2,163	5,017	12.94%	D
15	Banyumas	410	21	431	5.27%	579	180	759	1.96%	D
16	Kebumen	250	13	263	3.21%	693	29	722	1.86%	D
17	Wonogiri	201	40	241	2.95%	476	291	767	1.98%	D
18	Temanggung	164	14	178	2.18%	320	89	409	1.05%	D
19	Batang	128	49	177	2.16%	549	460	1,009	2.60%	D
20	Rembang	78	77	155	1.89%	122	843	965	2.49%	D
21	Pemalang	22	116	138	1.69%	44	582	626	1.61%	D
22	Cilacap	118	6	124	1.52%	335	66	401	1.03%	D
23	Wonosobo	97	14	111	1.36%	287	81	368	0.95%	D
24	Pati	50	37	87	1.06%	135	417	552	1.42%	D
25	Demak	31	41	72	0.88%	47	251	298	0.77%	D
26	Purbalingga	20	27	47	0.57%	106	208	314	0.81%	D
27	Banjarnegara	27	13	40	0.49%	54	120	174	0.45%	D
28	Pekalongan	27	---	27	0.33%	108	---	108	0.28%	D
29	Karangaanyar	25	---	25	0.31%	51	---	51	0.13%	D
30	Brebes	10	12	22	0.27%	40	137	177	0.46%	D
31	Tegal	21	---	21	0.26%	139	---	139	0.36%	D
32	Pekalongan	---	7	7	0.09%	---	36	36	0.09%	D, E
Central JAVA	32	6,830	1,353	8,183	100.00%	26,026	12,743	38,769	100.00%	
%	91.43%	83.47%	16.53%		100.00%	67.13%	32.87%		100.00%	

Source CEMSED (2002)

products at the time this research was conducted in 2002. Although, such behavior may have changed now with growing pressure from consumers for products produced under environmentally sustainable conditions.

Based upon the discussion above, the preliminary research findings did not seem to support the hypothesis that access to export markets would embed local firms into quasi-hierarchical relationships that could, in turn, drive economic or social upgrading—firms in the cluster survey did not indicate either a shift toward increased formal registration of companies, or more formal employment relationships, or improved environmental practices in the use of sustainably-grown timber.

It has been noted in other studies, citing the cases of countries such as Ireland, Malaysia, Mexico, the Philippines, South Africa and Singapore, that strengthened linkages between SMEs and their global buyers can be a powerful impetus for modernizing and dynamizing local industries, where countries upgraded their local productive capacities and enhanced their industrial performance by being integrated into global supply chains (UNIDO, 2002). Such studies emphasize, however, that such changes are not automatic. Rather, they emphasize that there is an important role to be played by strong government policy in this regard, as well as the need to form active partnerships among all key stakeholders including government, global firms, SMEs and their support agencies (UNCTAD, 2002). In this regard, such reports suggest that an important element, which appeared to be missing in the preliminary research findings from Indonesian wood furniture clusters, that there was room for more proactive policy support. While significant and in many ways quite effective, efforts were already being undertaken by a number of stakeholders in the private sector, the organized trade unions and the environmental sector; their combined impact still needed to be strengthened to bring changes that could be widespread and sustained. It should be noted that, at the time this research was conducted, important initiatives such as a multi-stakeholder forum had been established in Central Java Province, and had received the support of the provincial government, laying the groundwork for the types of institutional and policy support that had been lacking in earlier stages of globalization of the local woodworking sector.

2.5 Conclusions

We return to our original question of whether global markets offer opportunities for clusters in developing countries to upgrade economically and socially beyond price-based competition that compels many SMEs to pay low wages, avoid tax burdens through informality and squeeze profit margins, leaving limited capital available to make investments in improving machinery, working conditions and management practices.

This chapter has raised considerations regarding upgrading opportunities and constraints for SME suppliers to global buyers based upon preliminary research findings collected from 2002–2004 among wood furniture clusters in Indonesia. The fundamental recommendation that arose from this research was that Indonesian wood furniture clusters are under pressure to re-evaluate their current markets and consider diversifying into new markets.

The competitive crisis faced by this sector is driven largely (but not exclusively) by resource constraints caused by diminishing national sources of mahogany and teak, thereby forcing a restructuring among firms in this industry. Especially in resource-based industries, competitiveness requires not only firm-level improvements of manufacturing capability but also wise management of upstream linkages regarding the cost, quality and sustainable supply of raw materials. Firms that shift to lower-cost woods can choose to continue a strategy of low-cost, low-skill and low profit margins for less demanding clients and higher volume production both on domestic and export markets. However, firms that choose to continue working with mahogany and teak will be forced by higher costs to use these precious woods more efficiently, leading most likely to strategies of smaller scale, higher value-added production for niche markets. The latter strategy raises implications for the workforce, as higher-paid, skilled workers will be needed. Companies will need to invest in training and/or recruitment of more skilled production workers. Specialized skills such as product development and design will also be required. Managers are likely to need entrepreneurial talents in order to identify new clients, market their products more effectively and perhaps move into branding their product.

One way to approach the diversification of markets would be to undertake an evaluation of different market opportunities and challenges. Four market possibilities, and the respective opportunities and challenges of each, are summarized in a simple matrix seen in Table 2.3.

It is particularly interesting to explore the implications of a decision to shift into higher value-added niche markets, as this strategy involves the greatest potential opportunity for joining economic and social upgrading. First of all, such an upgrading strategy could forestall large job losses resulting from reduced furniture exports and declining competitiveness of a low-cost, low-wage production strategy. Second, the shift into higher value-added and more design-intensive products could open a new range of market opportunities. Third, higher value-added per unit of furniture would be based upon more efficient use of scarce timber. Fourth, this would also raise the need for worker skills, suggesting that current workers could retain their jobs, and could possibly raise their wage levels, by taking training courses and learning new skills. The emergence of new skills, such as design capabilities using computer-aided design might be a catalyst helping young people to find employment in areas of emerging and relatively scarce new skills. Greater demand for a more skilled and stable labor force may result, leading to possible improvements in conditions of work and hiring practices. Such an upgrading trajectory could eventually play a role in encouraging more stable contracts for workers and formal hiring of a greater segment of the labor force.

Table 2.3   Summary of main opportunities and challenges of four markets for wood furniture suppliers

Market	Opportunities	Challenges
Mass market global buyers and retailers	- Large orders - Process upgrading	- "Lock in", lack of functional upgrading opportunities - Risk of becoming low-wage, low-skill supplier - Constraints on raw material inputs, due to diminishing supplies of mahogany and teak
Niche markets for higher value-added wood furniture	- Functional upgrading - Reduced demand for scarce raw materials - Higher value-added per unit produced	- Adapting to smaller batch production - Identifying and contacting new "niche" buyers - Upgrading skills of labor force - Developing design skills
Domestic buyers	- Escapes fluctuating order volumes with global buyers - Can use cheaper, more common woods	- Less demanding market in terms of quality and delivery schedules - Potentially lower scale of orders
Importing Agents	- Access to global buyers that want more sophisticated products - Higher unit value than on domestic market	- Middle man takes a share of the profits

Perhaps this could lead toward registration of firms that currently operate in the informal economy.

Implications for institutions and policies also arise. For example, there is a need for SME support policies and credit facilities to alleviate the costs and risks of economic as well as social upgrading that may be prohibitive for local firms to undertake alone (and furthermore, this may be an incentive for informal firms to register, in order to qualify for such support). Furthermore, the prudent stewardship of natural resources is a task not only for individual firms, but for collective action, and must be tackled at different levels of government and with stakeholders at the sectoral level. As can be seen, an integrated upgrading strategy should join the main stakeholders (enterprises, workers and national and local governments, NGOs) to work together in addressing the complex range of competitive pressures that are bearing on this industry. This type of multi-stakeholder forum was formed in Central Java Province at the time this research was completed, providing an important mechanism to discuss problems and agree upon specific policy interventions that could be of use in the restructuring of the sector.

References

Amin, Ash and K. Robins (1990). "Industrial district sand regional development: limits and possibilities", in Pyke, F. et al, (eds) *Industrial Districts and Inter-Firm Co-Operation in Italy*, International Institute for Labour Studies, Geneva.

Becattini, G. (1990). "The Marshallian industrial district as a socio-economic notion", in Pyke, F. et al. (eds) *Industrial Districts and Inter-Firm Co-Operation in Italy*, International Institute for Labour Studies, Geneva.

Becattini, G. (2004). *Industrial Districts: a New Approach to Industrial Change.* Cheltenham, U.K.: Edward Elgar.

Bellandi, M. and A. Caloffi (2007). "District internationalization and trans-local development", Entrepreneurship and Regional Development, March.

Brusco, S. (1982). The Emilian model: Productive decentralisation and social integration. *Cambridge Journal of Economics*, 6, 167–84.

CEMSED — Center for Micro & Small Enterprise Dynamics (2002). Value Chain Analysis of Furniture Clusters in Central Java. Report on a survey prepared for the ILO, Salatiga.

Criscuolo, A. (2005). "Considerations for upgrading clusters of local enterprises in the global economy: the experience of the Third Italy", unpublished LED Working Paper, ILO, Geneva.

Gereffi, G. (1999). International trade and industrial upgrading in the apparel commodity chain. Journal of International Economics, 48(1), 37–70.

Gereffi, G. and Korzeniewicz, M. (eds) (1994). *Commodity Chains and Global Capitalism*. Westport, CT, USA: Praeger Press.

Gereffi, G. and Kaplinsky, R. (eds) (2001). The value of value chains: Spreading the gains from globalisation. *IDS Bulletin*, Institute of Development Studies, Vol.32, No.3, July.

Humphrey, J. (2003). Opportunities for SMEs in Developing Countries to Upgrade in a Global Economy. IFP/SEED Working Paper Number 43, ILO, Geneva.

Humphrey, J. and Schmitz, H. (2001). Governance in global value chains. *IDS Bulletin*, Institute of Development Studies, Vol.32, No.3, July.

Indonesian Central Bureau of Statistics, Jakarta (2002).

International Labour Office (2007). Green jobs: Climate change in the world of work. *World of Work Magazine*, no. 60, ILO Geneva, August.

Kaplinsky, R., Morris, M. and J, Readman (2002). The globalization of product markets and immiserising growth: Lessons from the South African furniture industry. *World Development*, 30(7), 1159–78.

Locke, Richard (1995). *Remaking the Italian Economy*. Ithaca, NY: Cornell University Press.

Nadvi, K. (1994). Industrial District Experiences in Developing Countries. In United Nations/GATE, *Technological Dynamism in Industrial Districts: An Alternative Approach to Industrialization in Developing Countries?* Geneva, Switzerland: United Nations/GATE, Geneva.

Piore, M. and C. Sabel (1984). *The Second Industrial Divide: Possibilities for Prosperity.* New York: Basic Books.

Piore, M. (1990). Responses to Amin and Robins. In F. Pyke, G. Becattini and W. Sengenberger (eds) *Industrial Districts and Inter-Firm Cooperation in Italy.* Geneva, Switzerland: International Institute for Labour Studies.

Pyke, F., Becattini, G. and Sengenberger, W. (eds) (1990). *Industrial Districts and Inter-Firm Co-Operation in Italy.* Geneva, Switzerland: International Institute for Labour Studies.

Readman, J. (2002). *Global Value Chains of Wood Products and Wood Furniture Industries.* Report prepared for ILO.

Sandee, H., Sulandjari, S., and Rupidara, Neil (2002). Business networks and value chains in furniture production : An analysis of demand-supply relationships of teak furniture from Central Java, Indonesia. First draft of a report to the ILO, Amsterdam, December.

Schiller, J. and Martin-Schiller, B. (1997). Market, culture and state in the emergence of an Indonesian export furniture industry. *Journal of Asian Business*, 13(1), 1–23.

Schmitz, H. (1989). Flexible specialisation: a new paradigm of small scale industriali-sation. *IDS Discussion Paper*, Number 161, University of Sussex, Brighton.

Schmitz, H. and B. Musyck (1994). Industrial districts in Europe: Policy lessons for developing countries? *World Development*, 22(6), 889–910.

Sengenberger, W. and F. Pyke (1992). *Industrial Districts and Local Economic Regeneration.* Geneva, Switzerland: International Institute of Labour Studies.

Tendler, J. (2002). Small firms, the informal sector, and the devil's deal. *IDS Bulletin*, Institute of Development Studies, Sussex, Summer.

Tissari, J. (2002). Potential and Status of Further Processing of Tropical Timber in the ASIA-Pacific Region. Paper presented to the Workshop on Further Processing of Tropical Timber in the Asia-Pacific Region, Korea Forest Research Institute, 9–12 July.

UN—United Nations Organisation (2007). *World Economic Situation and Prospects 2007.* New York: United Nations DESA.

UNCTAD (2000). *Handbook of Trade and Development Statistics.* New York: United Nations.

UNCTAD (2001). COMTRADE Data Base. New York: United Nations.

UNCTAD (2001). *World Investment Report.* Geneva: United Nations.

UNCTAD (2002). Improving the competitiveness of SMEs through enhancing productive capacity. Report by the UNCTAD Secretariat, Geneva, 20 December.

UNIDO (2002). *Industrial Development Report 2002/2003.* Vienna: United Nations.

Chapter 3

Environmental Upgrading of Industrial Clusters: Understanding their Connections with Global Chains in the Brazilian Furniture Sector

Jose Antonio Puppim de Oliveira[1]

3.1 Introduction[2]

There is a well-known, and recurring, myth that companies are closing down and jobs are being lost in developed countries because of the unfair competition from firms in less developed countries (LDCs) without environmental, labor, health and safety standards. However, many firms in developing countries are going beyond the "race to the bottom", and being competitive and at the same time searching for better environmental and labor standards and health-and-safety practices. Understanding how and under what conditions small firms in industrial clusters[3] in LDCs are upgrading their environmentally related standards is the objective of this chapter.

One explanation for social upgrading[4] is that clusters of Small and Medium Size Enterprises (SMEs) in developing countries have been increasingly connected

1   Development Planning Unit (DPU), University College London (UCL), UK, and Laboratorio do Territorio (LaboraTe), the University of Santiago de Compostela (Spain). During the edition of this book he was at the Brazilian School of Public and Business Administration (EBAPE) of the Getulio Vargas Foundation (FGV), Rio de Janeiro (Brazil).

2   This chapter is a result of a joint research project with Delane Botelho and José Jorge Abdalla of EBAPE/FGV. I thank Milber Fernandes for help in compiling the data. However, the author is solely responsible for all errors, opinions, analyses and arguments in the paper. I thank the EBAPE/FGV for their support through a research grant from the ProPesquisa.

3   There are many definitions of clusters (Porter, 1998; Schmitz, 1999; Markusen, 1996), but no consensual one (Martin and Sunley, 2003). I use a loose definition of cluster, as I did in Chapter 1. A cluster is an agglomeration of economic agents in the same economic sector within certain geographical boundaries. The agents in a cluster have actual or potential interactions with each other and with supporting organizations. Clusters may have medium and large firms.

4   Social upgrading is the improvement of social, environmental, labor and/or health-and-safety standards of a firm or group of firms, which may affect their surrounding communities.

to global chains. This has influenced the quality of their products and processes, as well as their environmental and social standards. The connection of those clusters to global chains can be a driving force to press for better standards, and help to facilitate processes of social and environmental upgrading both in the firms in the LDC and in the clients in distant markets.

In order to understand the relation between global chains, cluster development and upgrading of environmental standards, this chapter presents research based on a survey of 76 exporting firms of two furniture clusters in Southern Brazil: Bento Gonçalves (BG) in the state of Rio Grande do Sul (RS) and São Bento do Sul (SBS) in the state of Santa Catarina (SC). Those clusters are responsible for more than 70% of Brazilian exports of furniture and are connected to global chains in different ways.

3.2 Clusters, Value Chains and the Environment

Little research exists about how and why competitive clusters in LDCs are dealing successfully with their environmental and labor standards, and health-and-safety issues. Small firms are regarded as lacking the technical, institutional and financial resources for social upgrading. In many contexts of developing countries, there is a perception that if stringent regulations over those issues are enforced, the competitiveness of the firms/clusters would be undermined. Socially upgrading would be a burden, and governments should "protect" them from complying with those standards instead of enforcing the laws. Local authorities are inclined to make a devil's deal[5] to allow firms to remain with those low standards (Tendler, 2002). Sometimes they even give incentives for firms to stay informal. However, many clusters in LDCs have been able to socially upgrade and be competitive at the same time, and the external market forces have been an important pressure for change. Recently some authors have studied those issues (see Damiani, 2003; Tewari and Pillai, 2004; Samstad and Pipkin, 2005; Kennedy, 1999; Galli and Kucera, 2004).

There is a significant amount of literature on environmental management and global governance, as well as on the competitiveness of clustered firms in developing countries (Cassiolato and Szapiro, 2003; Schmitz and Nadvi, 1999; Mead and Liedholm, 1998), but they are not well-connected. The cluster literature has focused mostly on economic or production upgrading, giving little attention to environmental, labor or health-and-safety issues. The reason those topics are not common in the cluster literature may be because the early cluster studies, which

---

5   Tendler (2002): "a kind of unspoken deal between politicians and their constituents, many small-firm owners, including those in the informal sector: 'If you vote for me, according to this exchange, I won't collect taxes from you; I won't make you comply with environmental, or labor regulations; and I will keep the police and inspectors from harassing you.'"

helped to develop the cluster theories, were done in developed countries (e.g., USA and Italy), where laws exist and are more effectively enforced compared to developing countries. Therefore, environmental/labor/social issues were assumed to be enforced or were regarded as law enforcement or social policy issues; but implementation of public policies or enforcement of regulation over environmental or labor issues in developing countries still face tremendous obstacles (Puppim de Oliveira, 2002, 2008).

Moreover, the literature on cluster and global value chains are not well connected, even though they deal with similar objects. The literature on clusters focuses mostly on the internal dynamics to explain cluster governance and little on relations with global chains (Pyke, Becattini and Sengenberger, 1990; Humphrey and Schmitz, 2002). On the other hand, the global value chain literature pays little attention to the influence of the chain on the governance of clusters (Gereffi, Humphrey and Sturgeon, 2003). Table 3.1 gives an outlook of the debates over governance, relation to the external world, upgrading and competitiveness in the literature on clusters and value chains (Humphrey and Schmitz, 2002). An understanding of how value chains affect cluster governance would shed some light on how clusters can socially upgrade when they are linked to global value chains. Some studies have shown that demand (market) is an important agent for pushing firms to upgrade (Tendler and Amorim, 1996).

This chapter is an attempt to link this literature in order to explain the upgrading of clusters in the environmental sphere, as global chains seem important to push for improvements in environmental standards. However, my argument is that there is no linearity in the explanations of the upgrade of environmental standards and external markets. More exports do not always mean better environmental standards. Not only the intensity of exports, but the kind of insertion in the global chains (such as features of products and markets exported) and local factors (such as features and size of the firms) seem to be fundamental to determine environmental response of clusters.

Companies that want to increase resource productivity often need to have environmental demands on suppliers in the chain to reach their objectives (Esty and Porter, 1998). Initiatives of life-cycle analysis and management have also reached suppliers in the chain (Faruk et al., 2001), many of which are in the developing world. There is also a growing regulatory demand for Extended Producer Responsibility (EPR) in several developed countries, like those that are part of the European Union (Mayers, 2007). Those demands reach many suppliers in developing countries, whose environmental standards for their products are fundamental to plan for meeting the EPR regulations. Moreover, social and political organizations in developed countries, such as consumer groups and NGOs, have also pressed for social and environmental improvements in the chain. Finally, voluntary standards have affected the supply-chain (Rosen, Beckman and Bercovitz, 2002). Those standards can be international norms (e.g., ISO14001) or industry or company specific. For example, Motorola has extended its environmental standards to its subsidiaries and their suppliers leading to an improvement in their environmental quality (Rock, Angel and Lim, 2006).

As a result, companies have implemented many industrial ecology initiatives, both mandatory and voluntary; as production becomes more global, some of those initiatives reach firms in developing countries through their chains. Firms in LDCs try to respond to those external pressures by looking for alternatives working with other firms and actors in the clusters, sometimes with the help of external actors, including the buyers. For example, leather clusters in India had to improve their environmental standards because of demands from the German clients, who in turn provided support for environmental upgrading (Tewari and Pillai, 2004). In the furniture sector, environmental quality has become an important demand on product and processes, such as FSC and ISO 14001, but little is still researched.

This research tries to understand how those interactions have happened and how the different kind of links between clients and firms in certain clusters help to determine environmental changes in clusters. The next section analyzes how environmental demands have played out in the case of two furniture clusters in Southern Brazil. Firms in those clusters have received different kinds of pressure from global chains in order to improve its environmental standards.

Table 3.1   Debates in the literatures of clusters and value chains (adapted from Humphrey and Schmitz, 2002; Humphrey and Schmitz; 2000)

	Clusters	Value Chain
Governance within the locality	Local governance characterized by close inter-firm cooperation and active supporting organizations. Risks attenuated by local mechanisms for risk-sharing	Not discussed. Local inter-firm cooperation and government policy largely ignored
Relation with the external world	External relations not theorized, or assumed to be based on arm's length perfect market relations	Strong governance within the chain. International relations managed through inter-firm networks based on quasi-hierarchical relations. Risks attenuated by relationships within the chain
Upgrading	Emphasis on incremental upgrading (learning by doing) and the spread of innovations through interactions within the cluster. For discontinuous upgrading, local innovation centers play an important role	Incremental upgrading is possible through learning by doing and the lead of chain's leading firm. Discontinuous upgrading can be possible by entry into a different value chain
Key competitive challenge	Promoting collective efficiency through interactions within the cluster	Gaining access to chains and developing linkages with major customers

3.3 Furniture Industry in Brazil

The Brazilian furniture industry was composed of some 13,500 firms in 2001 (Coutinho, 2001), out of which approximately 10,000 are microenterprises (less than 15 employees) and only 500 or less are medium or large companies (more than 500 employees). Most of those firms are located in the southern and southeastern region of the country. The main clusters are Bento Gonçalves (state of Rio Grande do Sul), São Bento do Sul (state of Santa Catarina), Arapongas (state of Paraná), Mirassol, Votoporanga and São Paulo (state of São Paulo), Linhares (state of Espirito Santo) and Uba (state of Minas Gerais) (Coutinho et al., 2001; Leitura Moveleira, 1999; Souza, 2004). Figure 3.1 shows the main furniture clusters in Brazil.

According to the Brazilian Ministry of Development and Industry, the Brazilian furniture industry employs more than 300,000 workers directly and generates around 1.5 million indirect jobs in the chain. In 2005, the total revenues were estimated in R$ 12 billion (around US$ 6 billion). The production is composed of 60% of residential furniture, 25% of office furniture and 15% of others (furniture for schools, hospitals, hotels, etc). The sector has been very dynamic in the last

Figure 3.1   Location of Brazilian furniture clusters

years, even though the Brazilian economy has not grown much. Production of furniture grew more than 60% between 2000 and 2004, and exports almost doubled (Table 3.2).

In this research, we analyze the clusters around São Bento do Sul (SC) and Bento Gonçalves (RS), which are responsible for more than 70% of the Brazilian exports of furniture (Table 3.3). The cluster of Bento Gonçalves (number 2 in Figure 3.1), which also includes the municipalities of Flores da Cunha and Antonio Prado, has around 160 firms, employs more than 6,000 workers and generates annual revenues of around R$ 1 billion (US$ 500 million) in 2001. São Bento do Sul (number 12 in Figure 3.1) is the largest exporting cluster in Brazil (43% of the total exports) and is composed of 170 companies, including the municipalities of Rio Negrinho and Campo Alegre. The paper will focus on the analysis of environmental responses of those two clusters.

Table 3.2   General data on the furniture industry in Brazil (Abimovel, 2006)

Year	2000	2001	2002	2003	2004	2005
Revenues (R$)	7,599	8,631	10,095	10,756	12,543	12,051*
Exports in US$	485	479	533	662	941	991
Imports US$	113	99	78	70	92	108
Commercial Balance in US$	372	380	455	592	849	883
Exports/Production(%)	10.1	11.6	15.4	17.2	22.0	18.3*
Imports/Production (%)	2.5	2.6	2.6	2.3	2.6	2.3*

Source: Abimovel, 2006. Obs:* Estimates. R$ = Real (Brazilian currency, US$ 1 was equal to approximately R$ 2 in 2006).

Table 3.3   Main exporting states (in US$)

State	Exports (In US$)	(% of the Total Exports)
Santa Catarina	433,338,634	43,75%
Rio Grande Do Sul	270,442,545	27,31%
Parana	91,731,990	9,26%
São paulo	87,427,269	8,83%
Others	194,915,484	19,68%
Total	990.424.209	100,00%

Source: Abimovel, 2006.

3.4 Methodology

The research consisted of analyzing how and why two furniture clusters have responded to environmental pressures and what is the role of pressures from the global market chains. Those clusters were located in the regions of Bento Gonçalves (BG) in the state of Rio Grande do Sul and São Bento do Sul (SBS) in the state of Santa Catarina.

I chose the furniture sector in Brazil because it has made significant efforts for increasing its exports in the recent years. The sector is also interesting due to the different environmental aspects of its production. The production processes, the products and the inputs can be subjected to environmental demands. Environmental aspects of the production include painting methods, labor health-and-safety and waste. The demands on inputs and products can include wood certified by FSC (Forested Stewardship Council) and nontoxic paints. The two clusters were selected because they are the two largest exporters of furniture in Brazil and have different modes of production (SBS produces mostly furniture made of rigid wood for export, and BG manufactures primarily furniture made of MDF for the domestic market, but exports are strong as well).

The research consisted of three parts. First, several trade unions of the furniture sector were contacted to identify firms that have exported in the last three years.[6] A set of visits was made to SBS to understand the main environmental and export issues the sector has faced. Second, a survey was made among the 76 furniture industrial firms that export, covering more than 90% of the universe of exporting firms in both clusters. Out of those, 28 were located in Bento Gonçalves (BG) and 48 in São Bento do Sul (SBS). The results of the survey were collected personally by students of two local universities between December of 2005 and March 2006. The responders of the surveys were the people responsible for the exports sector of the firm.[7] Third, a series of 65 semi-structured interviews were made with some of the key actors of both clusters between September and November of 2006. Those interviews were made personally with two field visits and by phone with firm employees and owners, public officials, academics and members of civil society groups, trade unions and other supporting organizations in the two clusters.

---

6 The trade unions were the Sindicato das Indústrias de Construção e Mobiliário de São Bento do Sul (Sindusmobil), Sindicato das Indústrias de Construção e do Mobiliário de Bento Gonçalves (Sindmoveis), Associação das Indústrias de Móveis do Estado do Rio Grande do Sul (Movergs) and the Associação Brasileira das Indústrias do Mobiliário (Abimovel).

7 The person responsible for the exports varied a lot from firm to firm. In some firms, there were people responsible for dealing with the exports (international sales person), who dealt with agents and clients abroad. Others had only a general sales person or marketing personnel. In few small firms, the owner responded to the question.

3.5 Results and Analyses of the Survey and Field Research

The survey consisted of 13 questions with several items to detect the level of agreement using the Likert scale from 5 to 1 (5 = fully agree, 4 = agree, 3 = neither agree nor disagree, 2 = disagree and 1 = fully disagree). The results of the survey are presented in graphs in the figures below. For each question there is a graph of the average results of the exporting firms (all firms surveyed) and a second graph showing the average responses for each of the clusters separately, so it is possible to compare the answers. The survey is explained in more detail with information collected from the interviews.

3.5.1 Clusters are Different, so is the Pressure for Environmental Upgrade

Even though the clusters apparently look similar (large exports, high productivity and international competitiveness), they are very different in several aspects, including the environmental pressures and responses.

Companies have to deal with environmental requirements from different external actors in their daily operations. According to the survey, international clients are the main driving source of environmental standards on average (Figure 3.2), followed by the state environmental agencies and insurance companies. Communities and NGOs demand the least environmental standards from furniture companies. Banks and national clients also demand little environmental accountability from national clients. However, the main source of pressure for the two clusters is different. International clients and insurance companies are the highest demander for SBS, while the state environmental agency is the main demander for BG. One of the reasons is the main exporting market. Seventy-one percent of the firms in BG said Mercosul and Latin America were the main external markets as compared to only 2% in SBS, which has Europe and USA as their main exporting markets.

SBS specializes in furniture made of rigid wood from pinus. This process is highly intensive in labor. Almost all production is exported, mostly through brokers, as 75% of the companies in SBS mainly use Brazilian or foreign agents to export (compared to 50% of BG). In general, those agents come with a specified design and ask for low prices. Most of the furniture pieces are exported without local brand names, which receive the brand from the client. Basically, foreign clients outsource their production to SBS firms, which compete through prices with firms in Brazil and abroad. Their relationship with clients is that of quasi hierarchy as described in Table 1.1 (Chapter 1). This cluster follows the demands of international clients on environmental issues, which are high in their home countries. The need to avoid problems with their brands makes those clients demanding regarding environmental and social issues. Therefore, they are the main sources of pressure for environmental improvements in the firms in SBS (see Figure 3.2).

Firms in the BG cluster specialize in furniture made of flat wooden pieces, mostly MDF (medium-density fibreboard). The production processes in those firms are capital intensive, heavily automated and with low labor intensity. There is some production of furniture from rigid wood (pinus), but it represents less than 10% of the total production.[8] BG firms export, but their main focus is the domestic market of high quality and high-end furniture, such as modular kitchens for the middle class in the large urban centers. BG counts with strong brands of furniture in the Brazilian market, such as Todeschini and Dellano. The relation of the cluster with its clients is closer to the arm's length relationship in Table 1.1, but also some relations of quasi hierarchy. Firms follow the demands of environmental quality from international clients, but the most important pressure for environmental change is the state environmental agency (see Figure 3.2).

The level of environmental demand of the follow organizations is high

A. State environmental agency
B. Municipality
C. Banks
D. Insurance companies
E. National Clients
F. International clients
G. Trade union of the furniture sector
H. Community and Non-governmental Organizations (NGOs)

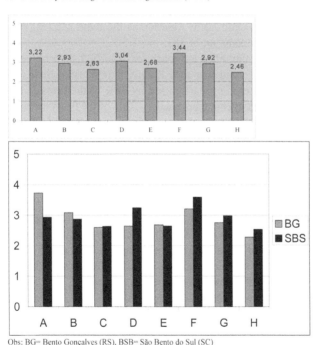

Obs: BG= Bento Gonçalves (RS), BSB= São Bento do Sul (SC)

Figure 3.2   Level of the environmental demand from different organizations

8   According to specialists in CETEMO (Center of Furniture Technology) in Bento Gonçalves.

3.5.2 Different External Markets and Marketing Approaches have Different Environmental Demands

Among the main demanders of environmental quality (international clients), the European Union is reported by far as the strictest, followed by the United States (Figure 3.3). Brazilian markets are the least strict environmental demanders, on the same level of Mercosul (Argentina, Uruguay and Paraguay).

The environmental demands of diverse external markets are different and receive different responses from the two clusters (Figure 3.3). As expected, both clusters perceive the European Union (EU) as the strictest market in terms of environmental quality, followed by the United States. Brazil and other countries in Mercosul have less strict environmental demands than those in more developed countries. SBS has a slightly higher perception of the environmental demands in all markets than BG. Countries in Mercosul (Argentina, Uruguay and Paraguay) are less rigorous than Brazil for firms in BG, while interviewees in SBS think otherwise. The difference seems to be due to the experience in the Brazilian and Mercosul market. BG has a larger experience in both markets, while SBS has less experience. BG has a strong position in domestic markets, as opposed to SBS, which focuses on exports. Moreover, 71% of the firms in BG have countries in Mercosul (20 out of 28) and Latin America as the main external market, while SBS has only 2%(1 company out of 48).

Regarding the main marketing approach for exporting, 46% of the respondents in BG and 29% in SBS said that they mainly export using their own contacts directly. The export agent is primarily used by 50% of the firms in BG and 75% in SBS. Brazilian export agents were used by 32% of the firms in BG and 52% in SBS. Foreign agents were the main channel for export of 18% in BG and 23% in SBS.

Demands are made directly by the final clients abroad (BG has more of those than SBS) or transmitted via export agents (SBS has more of these than BG). Those agents focus their demands on the features of the product, like certified wood or nontoxic paint. They also make their demands an explicit requirement of quality, and companies do not get higher prices if they perform better than agents ask. One entrepreneur said that those demands are part of the contract. If firms do not perform, they can have their contracts canceled and passed to other firms. Many of the firms have their production dependent on one or two agents, so they cannot lose any contract or will face significant economic problems. Therefore, firms take the demands very seriously, as they may not have another chance. Thus, those firms, mostly in SBS, perceive external environmental demands higher than in BG. Companies that export directly with their own brands (like many in BG) are more subject to the demands of the retails or consumers, which seem to be less intense than those of the agents. They can also make more mistakes, as they have more room to adapt. Noncompliance with environmental demands would not threaten the company because it has its own brand and would also have the option of a domestic market in case they lose a contract.

Environmental requirements in the following markets are high

A. Brazil
B. Mercosul
C. United States
D. European Union

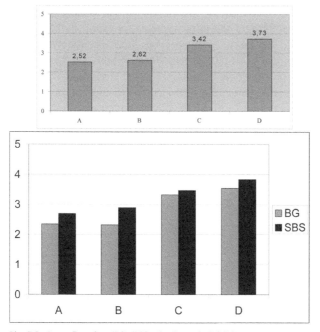

Obs: BG= Bento Gonçalves (RS), BSB= São Bento do Sul (SC)

Figure 3.3   Level of the environmental requirement from different markets

3.5.3 Environmental Quality is Perceived to be More Important for Export-Oriented Cluster

Regarding environmental management, companies agree that the external market is stricter than the Brazilian market and have invested in improving environmental standards (Figure 3.4). They also think that the FSC certification (for wood origin) is more important than the ISO14001 (environmental management system).

The survey pointed that most of firms (63 out of 76) had an environmental license from the state environmental agency. Wood certified by FSC certification was held by 43% of the companies, being more popular than the ISO14001, which only 9 companies held. This is in tune with Figure 3.4, as exporting companies mentioned that FSC was more important than ISO14001, especially in the exporting-oriented cluster of SBS.

Firms in the SBS perceive more strongly than BG the need to improve environmental quality and get environmental certifications (see Figure 3.4) because they are export-oriented and need agents for export. Even though both clusters agree in similar levels (Figure 3.4, C) that the external market is stricter than the domestic market, SBS firms have stronger assertions about the need for certifications (FSC, ISO 14001). They have also agreed more powerfully that they have invested in environmental improvements for export in the last years.

Firms in BG are more confident that their environmental standards are sufficient for reaching external markets. Because they tend to export directly and use their own brands, they are more aware of the environmental demands abroad. BG firms also have a stronghold in the domestic and Mercosul market, so they do not depend much on exports to developed countries and care less about their demands. SBS seemed to be less confident as they suffer more pressure from external markets. Firms do not have contact with the final clients because those

Agreement with the different environmental related issues (Likert scale)
A.   The certification ISO 14001 (Environmental Management System) is fundamental for exporting our products
B.   The certification FSC (Forest Stewardship Council) is fundamental for exporting our products
C.   The external market is stricter regarding environmental standards than Brazil
D.   Our company has invested significantly to improve environmental standards

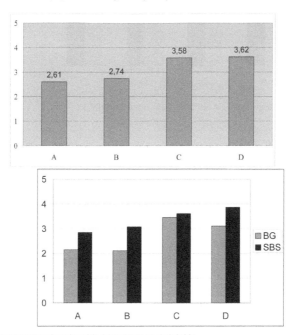

Obs: BG= Bento Gonçalves (RS), BSB= São Bento do Sul (SC)

Figure 3.4   Level of the importance of different environmental-related issues

clients purchase through the export agents. SBS entrepreneurs showed a high level of uncertainty about the final client demands because they feel the pressure, but they are not completely aware of the future changes. The certifications and investments are a guarantee for future demands. Under crescent pressures from the consumers, clients abroad also want more environmental guarantees from outsourced producers, if not they can move to outsource from other companies or regions (China, for example).

3.5.4 Clusters Perceive Different Kinds of Environmental Issues Differently

When the firm representatives were asked about the level of preparation their companies have to attend to environmental standards of external markets, they indicated that companies are more prepared to attend to labor health and safety standards, than environmental management systems and paint toxicity (Figure 3.5).

Both clusters perceive the different environmental demands of the external markets more or less evenly. The average response states that firms agree that they are prepared for all environmental demands asked in the survey (origins of inputs, paint toxicity, painting process, environmental management system and labor-related issues). São Bento do Sul (SBS) has a slightly higher confidence in their preparation than Bento Gonçalves (BG) except for labor-related issues.

The confidence of firms in SBS is based on the achievement of environmental quality asked by export traders in their contracts. Most of the managers interviewed mentioned that they rarely fail to comply with contract items related to environmental standards. As mentioned before, most of those standards are based on product quality (wood certification and paint).

The difference between the two clusters regarding preparedness for labor issues seems to be related to the kind of production they have. Firms in BG tend to be more capital intensive and automated. Their workers are less in number, but highly qualified. This makes BG less prone to informality and other labor issues like child labor and safety. Production in São Bento do Sul is more labor intensive. Even though labor is relatively well qualified and there are automated processes (but less than BG), there are a large number of small and informal firms that participate in the chain.

3.5.5 Cluster Factor Can Explain Environmental Responses

The representatives of the companies think that being part of a cluster helps the company to increase its exports and to a lesser extent to solve environmental problems (Figure 3.6). Firms in both clusters also tend to agree that being in a cluster, rather than isolated, has helped them to increase their exports and to a lesser extent to solve their environmental problems (Figure 3.6). Supporting organizations in both clusters, especially trade unions, have been very aggressive in trying to increase the exports of their members. Trade unions have made partnerships with the Ministry of Development and Commerce to give support to firms in order to

The firm is prepared to attend the following environmental standards of external markets

A.   Origin of the inputs
B.   Paint toxicity
C.   Painting process
D.   Environmental management system
E.   Labor health and safety

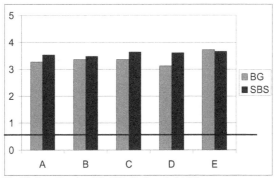

Obs: BG= Bento Gonçalves (RS), BSB= São Bento do Sul (SC)

Figure 3.5   Level of preparedness of some kinds of environmental-related demands

participate in international fairs. Many times the companies organize themselves and show up as one cluster, rather than individual firms. Before the 1990s, only large firms could afford to participate in those fairs. Also, both cities (Bento Gonçalves and São Bento do Sul) are famous for their annual international fair of furniture, which has attracted an increasing number of international clients and agents. Those actions have relatively succeeded in boosting export (as mentioned before, the two clusters account for more than 70% of the Brazilian exports of furniture). Firm managers mentioned the supporting organizations as the differential of being located in a cluster, such as trade unions and technical/training organizations. For example, the Center of Furniture Technology (CETEMO) of the National Service of Industrial Learning (SENAI) in BG is accredited to make several tests necessary for exporting for the main furniture markets, including an environmentally related one (e.g., presence of certain substances in the paints).

Being in a cluster is less advantageous for helping firms to solve their environmental problems than for exporting. Actually, this can even make environmental solutions more expensive. Firms in Bento Gonçalves and São Bento do Sul complained about the difficulty of finding a reliable supply of certified wood sometimes, because of the large demand of the cluster and small number of distributors. Indeed, wood wholesalers have a growing demand of certified wood from Chinese furniture companies, paradoxically the main competitors of the Brazilian firms in the most attractive markets of Europe and USA. Some

Benefits of being part of a furniture cluster

A. Being part of a cluster has helped the company to solve environmental problems
B. Being part of a cluster has helped the company to increase exports

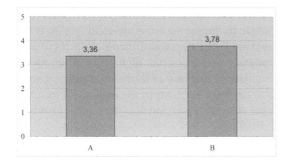

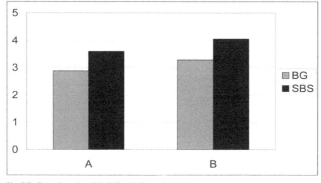

Obs: BG= Bento Gonçalves (RS), BSB= São Bento do Sul (SC)

Figure 3.6   Perceived benefits of being in a cluster

large firms, such as Todeschini in BG, have secured their own wood supplies by investing in pinus and eucalyptus plantation.

There is also a difference in the two clusters regarding the perception of the advantages of being in a cluster. Firms in SBS have a more positive view of being in a cluster than firms in BG. The reason for that is the larger presence of small firms in SBS. Many firms of BG are more self-sufficient, as they have their own resources such as technical capacity, suppliers, distributors and contacts. The cluster just provides them a supply of qualified labor, through the training centers in BG, like CETEMO (Center of Furniture Technology), presence of suppliers of inputs and connectedness with what is going on in the sector through the fairs. Supporting organizations make a difference being in a cluster. Few firms seem to be interacting with each other for creating collective efficiencies. Their most important relation is with the clients and suppliers, many of them outside the cluster. On the other hand, firms in SBS are mostly small and need the other firms and the supporting organizations in the cluster (trade union and SEBRAE) to help them to solve their bottlenecks. This is also perceived in the answers about the main obstacles for environmental upgrading (Figure 3.7). Firms in SBS were more prone to perceive obstacles to upgrading as compared to firms in BG.

Managers in the furniture companies feel that environmental information is the least problematic obstacle to improving environmental quality (Figure 3.7). The high costs of equipment are by far the most difficult obstacles to achieving better environmental performance, surpassing the high costs of specialized labor, the frequent changes in regulations and lack of financing. Joint projects in the cluster could somehow help to decrease the costs of the equipment for individual firms, such as the development of suitable equipment to the region, joint treatment plants or joint collection of residues.

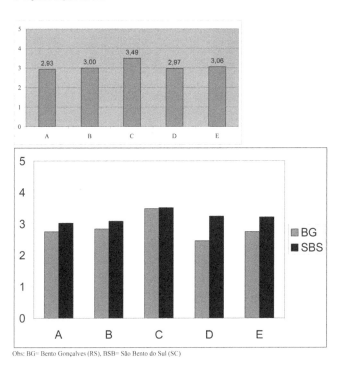

Obstacles to improvements in environmental quality

A. Lack of technical information
B. Frequent changes in environmental regulation
C. High cost of equipments
D. Lack of finance options
E. High costs of specialized labor

Obs: BG= Bento Gonçalves (RS), BSB= São Bento do Sul (SC)

Figure 3.7   Perceived obstacles to improvements in environmental quality

3.6 Final Remarks

Some companies have already mentioned environmental standards have prevented them from exporting in few occasions. Three companies affirmed that at least in one occasion in the last two years international clients refused to buy from them because of environmental standards. Clients in Canada and EU refused to trade because companies lacked certified wood and the European management system (EMA). Furniture companies surveyed have also started to demand environmental issues from their suppliers pushing demands down the furniture chain. Seventeen companies mentioned that they have asked environmental quality from their suppliers at least once. In most of the cases, the demands concerned certified wood and less toxic paints.

The insertion in global chains has put different demands on firms in LDCs to upgrade their environmental standards. Those demands vary according to the kind and intensity of insertion of clusters in global chains and local factors related to the features of the cluster. The two clusters studied in this chapter present different responses to environmental pressures from global chains and local actors. São Bento do Sul has been more sensitive to environmental pressures from external actors because most of its production is exported and based on quasi-hierarchical relations with buyers/agents. Bento Gonçalves is more confident of its environmental standards, as firms have a stronghold in domestic market depending less on international clients and depend less on agents to export. Moreover, the kind of market and the way firms are connected to them also influence response. BG firms focus mostly on direct sales for Latin America, and firms in SBS aim mainly at markets in the USA and EU through export agents, which pose a more explicit and direct threat to firms that do not comply with environmental quality demands, but at the same time does not reward companies that go beyond the expected quality (including environmental quality).

Firms in the cluster of SBS base their competitiveness on price and export through agents. Agents are important for allowing small firms to export and even pressing them to socially upgrade. However, they also leave little space for firm innovation beyond set quality standards, as firms do not get any reward for going beyond social or environmental quality of contract requirements. With the fall of the dollar in relation to the real (Brazilian currency), firms in SBS have had problems to compete with firms from China in the large markets like the USA and Europe. On the other hand, further improvements in environmental quality, which could differentiate them from the Chinese competitors, are not possible because agents do not look beyond contracts. Those environmental improvements have to come mainly through the enforcement of environmental regulations, which are loose in certain parts of developing countries.

References

ABIMOVEL — Associação Brasileira das Indústrias do Mobiliário (2006). *Panorama do Setor Moveleiro no Brasil*. São Paulo: ABIMOVEL.

Cassiolato, José Eduardo and Szapiro, Marina (2003). Uma caracterização de arranjos produtivos locais de micro e pequenas empresas. In: Helena M.M. Lastres, José E. Cassiolato and Maria Lúcia Maciel (eds), *Pequena Empresa: Cooperação e Desenvolvimento Local*. Rio de Janeiro: Relume Dumará Editora.

Coutinho, Luciano (2001). *Design na Indústria Brasileira de Móveis*. Curitiba: ABIMÓVEL Alternativa Editorial.

Damiani, Octavio (2003). Effects on employment, wages, and labor standards of non-traditional export crops in Northeast Brazil. *Latin American Research Review*, 38(1), 83–112.

Esty, Daniel C. and Porter, Michael E. (1998). Industrial ecology and competitiveness: Strategic implications for the firm. *Journal of Industrial Ecology*, 2(1), 35–43.

Faruk, Adam C., Lamming, Richard C., Cousins, Paul D. and Bowen, Frances E. (2001). Analyzing, mapping, and managing environmental impacts along supply chain. *Journal of Industrial Ecology*, 5(2), 13–36.

Galli, Rossana and Kucera, David (2004). Labor standards and informal employment in Latin America. *World Development*, 32(5), 809–28.

Gereffi, Gary; Humphrey, John and Sturgeon, Timothy (2003). The Governance of Global Value Chains, manuscript, (draft, later came in the Review of International Political Economy).

Humphrey, J., and Schmitz, H. (2000). Governance and upgrading: Linking industrial cluster and global value chain research. IDS Working Paper 120, Brighton, Institute of Development Studies, University of Sussex.

Humphrey, John, and Schmitz, Hubert (2002). How does insertion in global value chains affect upgrading in industrial clusters? *Regional Studies*, 36(9), 1017–27.

Kennedy, Lorraine (1999). Cooperating for Survival: Tannery Pollution and Joint Action in the Palar Valley (India). *World Development*, 27(9),1673–91.

Leitura Moveleira (1999). *O Mercado Norte-americano de Móveis*. Curitba: Alternativa Editorial.

Markusen, Ann (1996). Sticky places in slippery space: A typology of industrial districts. *Economic Geography*, 72, 293–313.

Martin, Ron and Sunley, Peter (2003). Deconstructing clusters: chaotic concept or policy panacea? *Journal of Economic Geography*, 3(1), 5–35.

Mayers, C. Kieren (2007). Strategic, financial, and design implications of extended producer responsibility in Europe: A producer case study. *Journal of Industrial Ecology*, 11(3), 113–31.

Mead, Donald C. and Liedholm, Carl (1998). The dynamics of micro and small enterprises in developing countries. *World Development*, 26(1), 61–74.

Porter, Michael E. (1998). Clusters and the new economics of competition. *Harvard Business Review,* 76(6), 77–90.

Puppim de Oliveira, Jose A. (2008). *Implementation of Environmental Policies in Developing Countries: A Case of Protected Areas and Tourism in Brazil.* Albany, NY, USA: State University of New York Press (SUNY).

Puppim de Oliveira, Jose A. (2002). Implementing environmental policies in developing countries through decentralization: The case of protected areas in Bahia, Brazil. *World Development,* 30(10), 1713–36.

Pyke, F., Becatttini, G. & Sengenberger, W. (1990). Industrial districts and inter-firm co-operation in Italy. Geneva: International Institute for Labour Studies.

Rock, M. T.; Angel, D. P. and Lim, Pao Li (2006). Impact of firm-based environmental standards on subsidiaries and their suppliers: Evidence from Motorola-Penang. *Journal of Industrial Ecology,* 10(1–2), 257–78.

Rosen, Christine Meisner; Beckman, Sara L. and Bercovitz, Janet (2002). The role of voluntary industry standards in environmental supply-chain management: An institutional economics perspective. *Journal of Industrial Ecology,* 6(3–4), 103–23.

Samstad, James G., and Pipkin, Seth (2005). Bringing the firm back in: Local decision making and human capital development in Mexico's Maquiladora sector. *World Development,* 33(5), 805–22.

Schmitz, H. and Nadvi, K. (1999). Clustering and industrialization: Introduction. *World Development,* 27(9), 1503–14.

Souza, M.B. (2004). Clusters como Estratégia de Desenvolvimento Sustentável de Pequenas Empresas do S0etor Moveleiro de Colatina. Master Thesis. EBAPE/FGV.

Tendler, Judith (2002). Small firms, the informal sector, and the devil's deal. *IDS Bulletin* [Institute of Development Studies] 33(3), July.

Tendler, J. and Amorim, A. M. (1996). Small firms and their helpers: Lessons and demand. *World Development,* 24(3), 407–426.

Tewari, Meenu, and Pillai, Poonam (2004). Global standards and environmental compliance in the Indian leather industry. *Oxford Development Studies,* 33(2), 245–67.

Chapter 4

Social Upgrading in Agriculture-based Clusters: Common Lessons from Cases in Asia and Latin America

Octavio Damiani[1]

4.1 Introduction[2]

A great deal of the literature on clusters focuses on industrial firms and organizations, analyzing the role of their agglomeration and local linkages among them in their competitiveness, learning, and upgrading. In contrast to industrial clusters, relatively low attention has been paid to agriculture-based clusters, defined as agglomerations of agroprocessing industries, farmers supplying raw materials for those industries, and industries and services serving farmers and industries. Farmers producing certain types of crops tend to agglomerate naturally as a result of soil types and climate conditions that generate high potential for growing such crops. Agroprocessing industries tend to establish in areas where there are agglomerations of agricultural producers in order to reduce transportation costs that characterize many agricultural products that are bulky or due to the loss of quality if raw materials had to be transported far away. Agroprocessing industries frequently enter into contract relations with the farmers supplying raw materials— a relationship that frequently has not worked well and thus has generated frequent debates among analysts and policymakers about their social impacts and if it could and should be supported as a part of a rural development strategy.

----

1   Octavio Damiani is an agricultural economist who graduated from the Universidad de la Republica (Uruguay), receiving Master's and Doctoral degrees from the Massachusetts Institute of Technology (MIT). He has worked extensively in policy studies and in the evaluation of, and technical assistance to, the implementation of rural development projects for several international organizations, including the World Bank, the Asian Development Bank, the Inter-American Development Bank, the International Fund for Agricultural Development, and the Food and Agriculture Organization.

2   This chapter is based on information collected as part of two separate evaluation studies carried out by the author for the Asian Development Bank and the International Fund for Agricultural Development, focused on the poverty exit strategies of rural households and on the development of organic agriculture among small farmers respectively. All views and interpretations presented here are those of the author and do not necessarily represent those of the organizations above. I am thankful to José Puppim for his encouragement to prepare this chapter and for the useful comments provided. All possible errors, however, are the sole responsibility of the author.

This chapter focuses on social upgrading in agriculture-based clusters, aiming at contributing to the understanding of how agglomerations of small- and medium-sized farmers and agroprocessing industries become competitive, while at the same time improve social, environmental and labor standards. It uses specific cases of agriculture-based clusters to analyze the relationship between the particular characteristics of the product markets, production upgrading, and poverty reduction. In addition, it analyzes the relationship between the clusters' governance and social upgrading, in particular the role of central and local governments, private companies, and farmer associations in the development and particular characteristics of the clusters.

The cases addressed in the study include the tobacco cluster in the southern portion of the Yunnan Province in the People's Republic of China and organic coffee production in the State of Chiapas in Mexico. While the two clusters have developed in very different political, economic and social contexts, they have in common the fact of having been successful in regions that were among the poorest in the respective countries. Although they were characterized by high levels of rural poverty, both Yunnan and Chiapas had excellent conditions in the natural environment to produce certain value crops, and in both cases, the development of the agriculture-based clusters involving high-value crops that are studied here played an important role in the reduction of rural poverty. Poor farmers were able to make substantial changes in their traditional production activities, introducing new products (high quality tobacco in Yunnan, organic coffee in Chiapas) characterized by high demand and particular product characteristics. Thus, changes in production and social impacts in both cases were driven by new linkages of the involved small farmers with production chains and markets that were demanding in product quality and had an interest in the social and environmental features of the production process, paying premium prices for these attributes and helping them in many ways to improve the production process.

The chapter is organized as follows. After this introduction, Section 2 describes the main characteristics of the two clusters and analyzes the relevance of the clusters in poverty reduction. Section 3 analyzes the main factors leading to social upgrading. Section 4 presents conclusions and policy implications that emerge from the cases.

4.2 The Cases

The case studies are based on fieldwork and desk review of background information. In the Yunnan Province, fieldwork was carried out in July 2005 as a part of a wider evaluation study on rural households' strategies to exit out of poverty implemented by the Operations Evaluation Department of the Asian Development Bank between 2004 and 2005, which also covered the Sichuan Province and other

Asian countries (Vietnam and Malaysia).[3] In Chiapas, fieldwork was carried out as a part of an evaluation study on the impacts of the adoption of organic agriculture among small farmers implemented by the International Fund for Agricultural Development (IFAD) between 2001 and 2002, which also covered other cases in Argentina, Dominican Republic, El Salvador, Guatemala and Mexico.[4]

In both cases, fieldwork was based on qualitative methods, including interviews to rural households, leaders of rural villages and grassroots organizations, representatives of firms' and workers' organizations, owners, managers and employees of agricultural firms and nonagricultural enterprises, policy makers and technical staff at government agencies both at national and local levels, and politicians at central and local levels. All the interviews were open-ended, having the objective of understanding the households' livelihood strategies and the role played by the introduction of the products studied here (tobacco and organic coffee) in their income and quality of life. Most of these interviews included field visits to the crops and farm facilities.

4.2.1 Tobacco Cluster in Southern Yunnan

Organization of the cluster   Yunnan is one of the five southern border provinces of China, neighboring Vietnam, Laos, and Burma. With a GDP per capita of CNY 4,840 (USD 583) in 2001, Yunnan ranked 28th out of the 31 provinces of China, being one of the poorest and least industrialized, with an average GDP and rural income per capita that are two-thirds of the national average. Yunnan is characterized by a great diversity of topography and climate, going from alpine mountains in the north to hills in the south. The total population is over 41 million, with 24 ethnic minorities reaching about 13 million (32% of the population) and dominating in rural areas. While Yunnan accounts for about 3.4% of China's population, its share in national GDP is only about 2.2%. The average annual GDP per capita growth rate for the whole Yunnan province during 1999–2003 was 6.2%. The share of agriculture in GDP in 2003 was 21%, falling from 42% at the end of the 1970s.

The tobacco industry is the most important sector in Yunnan, accounting for as much as 30% of China's output of flue-cured tobacco and 18% of the country's output of rolled cigarettes. While there are records of ancient tobacco plantations (as old as 1,700 years ago), modern tobacco production started at the time of World War II, using varieties imported from the United States.[5] In the early 1950s, Yunnan tobacco had become well-known for its excellent quality, being considered among the best in China. As a result, many cigarette factories in China started to purchase

---

3   Results of this study are presented in ADB (2005) and can be found in http://www.adb.org/Documents/SES/REG/Rural-Poverty-Targeting/ses-rpt.pdf. Also see Damiani (2006).

4   The results of this study are presented in Damiani (2001 and 2002).

5   See Maeda et al. (2003).

tobacco from Yunnan, which led to a rapid growth in the area and in production. However, all tobacco was industrialized in other provinces of China.

The start of the tobacco industry development in Yunnan took place in the 1970s, when new cigarette factories were constructed in the main tobacco producing areas following the setting up of tobacco curing plants. By the end of the 1970s, there were nine tobacco plants in Yunnan with a total of more than 10,000 employees. However, quality was very uneven and most of the production of cured tobacco was sent to be industrialized outside Yunnan, with Yunnan's cigarettes output reaching only 6% of the total production of cigarettes in China.

By the late 1990s, the Yunnan Province had become the most important producer of tobacco and one of the main producers of cigarettes in China. Seven of the top 10 enterprises of the Chinese tobacco industry in terms of the quantity of cigarettes inter-provincially allocated for monopolistic sales by the central government in 2000 were Yunnan-based enterprises. The tobacco cluster was organized around government-owned tobacco processing companies, among them the Yuxi Hongta Tobacco Group Ltd, which was part of a larger conglomerate (the Hongta Group) and the most important tobacco processing firm in the Yunnan Province. The Yuxi Hongta Tobacco Group normally operated signing contracts with individual small farmers, providing them with technical assistance and inputs and purchasing all their production.

Until the mid-1990s, the suppliers of Yuxi Hongta Tobacco were small- and medium-sized farmers in counties around the Yuxi county in the north of Yunnan, where the company had its main processing facilities.[6] At that time, the company decided to shift its production base to counties located in the south of Yunnan, mainly villages in the municipalities of Simao and Meng Lian. Partly as a result of the upgrading of an expressway that linked Yuanjiang and Mohei (a 147km section that was part of the road connecting Kunming, capital city of Yunnan, and Thailand), transportation costs decreased substantially, making it easier for farmers to sell in Kunming due to the decrease in transportation costs. A large proportion of the tobacco producers in northern counties of Yunnan decided to shift from tobacco to other crops of higher value, especially flowers, which they sold in Kunming.

The reduction in the area's production of tobacco made the Yuxi Hongta Tobacco Group search for new production areas. After carrying out some studies, the company identified that climate and soil conditions in mountain areas in the south of the Yunnan were exceptionally good for tobacco, so it decided to expand the tobacco cultivation there based on similar contractual arrangements with small farmers. Under the contracts, the firm provided seeds, technical assistance, training and financial support for building simple on-farm facilities for drying the tobacco

---

6   Besides tobacco manufacturing, the Hongta Group has diversified into other industries, including energy and transportation, banking and finance, insurance, real estate, medicine, and light chemical.

leaves. The inputs provided by the firms were deducted from the payment due to farmers at the time of the harvest.

Production and social changes brought by tobacco   The growth of tobacco in the south of Yunnan generated substantial changes in family-based agriculture. Until the mid-1990s, rural households in southern Yunnan had traditional crops and animal products as the major source of income, growing rice and corn and having small livestock (chicken, ducks, and pigs); all of them mainly for the households' consumption. By 2006, these production patterns had changed dramatically. Most farmers had been able to start growing tobacco, usually maintaining the production of paddy rice under irrigation and growing tobacco in dryland areas. A low proportion had started to cultivate tobacco under rotation with wheat—a production system aimed at avoiding the negative effects of monoculture. All production was purchased by the Yuxi Hongta Tobacco Group.

The cultivation of tobacco was usually accompanied by intensification and higher productivity. Both recent and older growers of tobacco had to make progressive adjustments to their technology that implied: a) intensification in the use of the land (more labor and more inputs per unit of land); b) a more timely use of fertilizers and pesticides; and c) the use of a new variety of tobacco. The timely use of pesticides was the result of efforts from the company to increase productivity and quality, and most households were slowly doing so.

The cultivation of tobacco had substantial positive impacts on poverty reduction, as it provided a new source of cash income to households that have few sources of cash, as their traditional production was mainly for family consumption. According to official statistics, rural poverty in Yunnan reached approximately 23% in 1997, falling from 41% in 1985. [7] In the counties in the southern part of the province that were included in the case study (Yuanjiang, Mojiang, Simao, and Meng Lian), poverty was higher than 30%, but it had also decreased in a similar way. Household income levels within villages were quite homogeneous, but the variation between villages was high, with the poorest being located in more remote and mountainous areas with poor roads. In some of the villages visited, the average income of poor households was CNY 625 or US$ 77 per year, while in others it was CNY 865 or US$ 106 per year, which is low even if compared with the official poverty line of US$ 0.66 per day used in China. Using the definition of poverty used by the rural households themselves (based not only on enough food to eat but insufficient cash income to cover other needs considered basic),

---

7   Official estimates of poverty in China are based on the government's austere poverty line equivalent to US$0.66 per day. The World Bank has developed an international poverty standard of US$1 per day (in 1985 purchasing power parity dollars) for cross-country comparisons. Estimates based on the international poverty standard of US$1 per day indicate substantially greater numbers of absolute poor both in Yunnan and in China as a whole, but confirm the continuing remarkable decline in poverty.

the proportion of poor households varied greatly between almost no poor in the better-off villages to close to all in the poorest ones.

Most households interviewed in the tobacco producing villages had been able to exit out of poverty based both on their own definitions of poverty and on the official poverty line. Most (including also those who were still poor) responded that their life was better than 5 or 10 years earlier.[8] Food shortages did not affect even the poor households, though their consumption of meat was low and some said that they occasionally experienced periods with difficulties to access enough food. Poor households did lack cash income to satisfy other needs, such as clothes or bringing a sick family member to the doctor. About half of the farmers had been able to accumulate some physical capital, including mainly drying rooms for tobacco and one or two buffaloes to help work the land. About half of those interviewed had their own drying houses—a higher proportion in those villages that had started earlier with tobacco cultivation. Drying houses are very important because fresh leaves are highly perishable and lose quality very quickly after the harvest, so they need to be dried. Thus, it was one of the first investments made by households once they had some resources available. The cost of both infrastructure and the equipment was subsidized by the tobacco company. Some households had also purchased other goods, such as motorbikes.

In the villages visited as a part of the study, the role of tobacco cultivation in poverty reduction becomes even more important because the vast majority of rural households did not consider migrating to urban areas as one of the possible livelihood strategies. In fact, migration to urban areas became the main strategy that rural households adopted to exit out of poverty in most provinces of China, with the younger household members usually migrating to cities in Eastern China to work in the industry and then sending back remittances, while also relieving the rest of the family of their burden.[9] In contrast, migration of rural population to other provinces was a rare phenomenon in southern Yunnan, mainly due to very low education levels and language barriers. Most villagers were of ethnic minority groups, had completed only a few years of primary school, spoke little or no Mandarin, and had no social connections with other provinces because there was no history of migration. Thus, most interviewees feared migrating to other provinces because they did not know anyone and had heard many stories of people not receiving their payment from their employers. Therefore, rural households in the tobacco producing villages studied here did not have migration as one of their possible strategies to exit out of poverty, which made agricultural diversification and intensification one of the main strategies that they had available.

Finally, the Hongta Tobacco Group had good connections with local governments in Yunnan and was able to obtain their support to construct and improve roads as a condition to expand the cultivation of tobacco in counties of the southern portion of the province. The construction and upgrading of rural roads also played a great

---

8   See ADB (2005).
9   See ADB (2005) and Damiani (2006).

role in improving the quality of life of tobacco producing communities, not only making tobacco cultivation possible and reducing the transportation costs of inputs and products, but also facilitating the access of households to social services (such as health and education), reducing the transportation costs of people, increasing the availability of food and other goods, and reducing the prices at the village level.

4.2.2 Organic Coffee in Chiapas

Organization of the cluster With a large indigenous population that concentrates in rural areas, Chiapas is one of the poorest states in Mexico. Poverty and marginalization were even in the roots of an uprising that took place in the mid-1990s led by the "Zapatista movement". Coffee has been a traditional crop among small indigenous farmers, who used to sell their production to middlemen and received low prices.

In the 1990s, the Soconusco region in Chiapas had developed into an important producer of organic coffee. The organic coffee cluster is organized around a farmer association named "Indígenas de la Sierra Madre de Motozintla San Isidro Labrador" (ISMAM). ISMAM is a well-known success story in Mexico, which by 2001 included 1,300 indigenous producers from 146 communities in 18 municipalities of the Soconusco region, with 200 others who were in the process of becoming members. All members of ISMAM are small farmers and cultivate only organic crops. These farmers cultivated close to 5,000 hectares of certified organic coffee—about 10% of the cultivated area of certified organic coffee in Mexico. ISMAM was created in 1988 with the support of the Catholic Church, which had long been working with indigenous communities in Chiapas. The newly created association was aimed at opening alternative marketing channels for the coffee production and avoiding middlemen.

The creation of ISMAM had its origins in a meeting of indigenous leaders in the Soconusco region, close to the border with Guatemala in the state of Chiapas, which the Church promoted in 1988. The Catholic Church had been working with indigenous communities in Chiapas for a long time, not only in religious activities, but also promoting changes and improvements in the production of coffee and other economic activities. The Church promoted the meeting of 1988 to discuss the crucial problems affecting coffee—the main crop in the region. Community leaders identified and discussed several problems, but they focused mainly on those related to the marketing and technology of production. The marketing of coffee was viewed as dominated by middlemen who paid low prices and did not care about the quality of the product. Meanwhile, the technologies promoted by the public extension services and input suppliers were viewed as inappropriate because they were based on expensive inputs that small farmers found difficult to purchase. The meeting concluded that the organization of small farmers was key to solve marketing and technology problems, and that it was essential to search for alternative sustainable technologies.The discussions in the meeting led to the creation of ISMAM at the end of 1988. The organization initially comprised 200 small indigenous farmers, and initially focused on the collective marketing of production and the strengthening of the organization at the grassroots level through two types of actions: a) the strengthening of collective work in order to reduce the need for hiring paid labor and reduce costs; and b) the creation of local committees to facilitate the participation of members in the decisions of the newly created organization.

At that time, the demand for organic products in European countries—with whom ISMAM had close links through coffee buyers and European Non-Governmental Organizations (NGOs)—had started to grow rapidly. Because most producers had a history of little or no use of chemical inputs, ISMAM's leaders saw a great opportunity in certifying the coffee crops of its members as organic, and they started efforts to focus their efforts on making the changes at the organization's and the farmers' levels required to obtain the organic certification. The first contacts with foreign certification agencies were established in the late 1980s through one of ISMAM's buyers of coffee. After initial inspections, the organization started a transition period of three years and was eventually successful in certifying the coffee production of its members in 1993. After a first year of selling its organic coffee through a farmer association in Oaxaca, ISMAM established its own marketing contacts and began exporting directly an annual average of 65,000 quintals (2,900 tons) of coffee mainly to Europe and the United States.[10] ISMAM was able to export its organic coffee at prices between 30% and 87% higher than the conventional coffee in the period 1993 and 2001. Since the mid-1990s, ISMAM was also able to sell about 30% of its coffee in the fair trade market, thus obtaining even higher prices—USD 165 per quintal in 2001, as compared with USD 75 per quintal of organic coffee and USD 40 per quintal of conventional coffee.

By 2001, ISMAM had 1,300 members in 146 communities in 18 municipalities of the Soconusco region in Chiapas, In fact, every member of ISMAM had a maximum of 10 hectares of coffee and had to produce all his crops organically, not being allowed a mix of organic and conventional products. The total area of ISMAM's organic certified coffee was 5,000 hectares, which represented an average of 3.8 hectares per farmer, being certified by recognized European and US certification agencies, including Naturland, Oregon Tilth Certification Organic (OTCO), and OCIA.. ISMAM exported about 65,000 quintals (close to 2,900 tons) a year to several countries, including Germany, Italy, France, Netherlands, Austria, Spain, the United States, Japan and Argentina. The high quality of ISMAM's coffee led to its recognition through several awards, including the National Exporters Award and a recognition from the Government of Chiapas in 1995, and the International Award for the Best Trademark by the International Trade Association in 1998.

------

10   One quintal = 100 lbs. = 44.5 kilograms.

ISMAM focused on the collective processing and marketing of coffee produced by their members, having a toasting and packing plant in the city of Tapachula with new equipment of 2.5 tons per hour processing capacity. In addition, ISMAM provided its members with technical assistance, training and credit. Technical assistance and credit focused on the improvement and monitoring of the application of organic technologies in coffee. In addition, it implemented a program to promote the diversification of its members' production, supporting the introduction and improvement of the production of organic honey (certified since 1996), and helping women to carry out alternative activities like husbandry and the cultivation of vegetables for family consumption. ISMAM also had a farm with 237 hectares of coffee.

Production and social changes brought by organic coffee Organic agriculture employs cultural and biological practices to control pests, the use of crop rotation to maintain soil fertility, and the application of animal and green manures instead of chemical fertilizers, virtually prohibiting synthetic chemicals in crop production and antibiotics or hormones in livestock production.[11] Thus, coffee producers in Chiapas had to introduce some technology changes in order to comply with the organic standards. Because they used to cultivate coffee under the shade of forests, applied labor-intensive technologies, and did not use chemical inputs (or used them only occasionally), they found it relatively easy to shift to organic production, having to make relatively small investments and minor changes in the technologies of production. The most important problem that they faced — a disease caused by a fungus called *Broca* — was controlled using manual practices rather than the chemical inputs that were characteristic among large farmers. The most important changes that ISMAM farmers undertook during the shift to organic coffee production consisted of the application of soil-conservation measures and of special management practices, such as the introduction of new species of trees to provide shade to the coffee plants. These new practices had relatively little investment costs and demanded mainly labor for the construction of terraces and other soil-conservation measures, which were the most costly of the changes.

In addition, the organic models of production have also been associated with positive effects in terms of the health of producers and workers and in terms of the environment. This statement, however, is based on qualitative evidence, as no measurements could be obtained to support it precisely. Most organic producers

---

11   According to the Codex Alimentarus, "Organic agriculture is a holistic production management system which promotes and enhances agro-ecosystem health, including biodiversity, biological cycles and soil biological activity. It emphasizes the use of management practices in preference to the use of off-farm inputs, taking into account that regional conditions require locally adapted systems. This is accomplished by using, where possible, agronomic, biological and mechanical methods, as opposed to using synthetic materials, to fulfill any specific function within the system." See FAO/ITC/CTA (2001), Chapter 1.

argue that their concerns about the potential ill effects of chemical inputs on health were an important factor in their shift to organic methods of production.[12]

As a result of these changes, yields of coffee increased about 50% in most ISMAM's farms during a time period of eight to ten years. Organic production had significant positive effects on farmers' incomes and quality of life. The successful marketing of coffee allowed ISMAM's members to receive higher prices for their coffee ($800 per quintal in 2000, 45% higher than the $550 received by conventional small producers, and $650 per quintal in 2001, 62.5% higher than the $400 received by conventional small producers).[13] In addition, the organic models of production had positive effects on the environment, as coffee was produced using environmentally-friendly technologies, under the shade of native trees, using little amounts or no chemical inputs, and applying soil conservation measures that were unusual among conventional coffee producers.

4.3 Key Factors in Production and Social Upgrading

This section analyzes factors that played a key role in production and social upgrading, namely global standards that had to be met by agricultural producers in order to be able to access markets, agroprocessing firms that produced under contract relations with small farmers, and small farmer associations that organized production and carried out monitoring systems to ensure farmers' compliance with organic methods of production.

4.3.1 Global Standards and Production Patterns among Small Farmers

Several authors have stressed the relationship between global standards and the production patterns in producing regions. In Chiapas, global standards of organic production (the main market for coffee producers) played a key role in the technology and social characteristics of coffee production. The shift from conventional to organic production of coffee can be explained by the dramatic growth in the international demand for organic products, especially in the European Union (EU), the United States, and Japan, that took place during the 1990s. Consumers in these countries have become more concerned about the effects of different types of food on health, the potential risks of exposure to pesticide residues in foods, and the effects of different production systems on the environment. This led to a dramatic growth in the consumption of organic products.

Organic production requires the application of specific production norms and certification procedures. The countries of the European Union endorsed a common organic standard in the early nineties, and Canada, Japan and the United States adopted organic standards and regulations in the late 1990s. In 1999, the

---

12   Damiani (2002).
13   Damiani (2002).

Committee on Food Labeling of the FAO/WHO Codex Alimentarus adopted guidelines for the production, processing, labeling and marketing of organically produced foods.[14] In all cases, product certification became a major issue, being intended to provide consumers with an assurance that certain standards have been met in the production process. Certification focuses on the materials and processes that the producers have used in the production of specific crops or animals. Organic production must be based on natural inputs, as certification prohibits the use of synthetic inputs. In general, certification standards also include the mandatory use of methods that contribute to maintaining or enhancing soil fertility. Some social issues are also considered, including not using child labor.

In practice, obtaining organic certification has become an essential procedure if an organic producer wishes to be able to sell products as organic. First, consumers tend not to trust claims that products are 'organic' if the products have not been certified. Second, most laws and regulations on organic agriculture elsewhere prohibit the use of terms such as 'organic' or 'natural' for products that have not been certified. Thus, a product that has not been certified has to be sold as conventional even if it has been produced according to all the specifications of organic production.

The organic certification is provided by specialized agencies, most of which are based in industrialized countries. The certification process starts with an application by a producer or a group of producers to a certification firm. The certification firm usually sends an inspector, who visits the production sites and determines if the production process meets "organic" standards. The inspectors do this based on interviews with producers, field visits to the croplands involved, reviews of the organic fertilizers and other inputs used and laboratory tests of samples of the soil, water and agricultural products. Some of the main requirements that must be met in order to obtain certification are: (a) the land under organic production must not have been used for conventional agriculture relying on chemical or synthetic inputs for a minimum time period (usually three years); (b) conventionally grown crops must be a minimum distance from the organic crops, and a forested area may be required as a barrier between the organically and the conventionally grown crops; (c) the inputs used in the production process must be organic, and no chemical or synthetic inputs are permitted; (d) soil-conservation measures must be applied and (e) small farmer cooperatives and other forms of associations must demonstrate that they are able to organize their own supervision system to ensure that organic standards are met by all members. Once the organic certification has been approved, it is valid for a one-year period, during which inspectors visit the sites usually twice without notice.

In the case of tobacco production in Yunnan, global standards also played a key role in production standards. Until the 1990s, cured tobacco production in Yunnan depended mainly on the good potential of its natural environment and the traditional experience accumulated by tobacco farmers. However, productivity

---

14   See FAO/ITC/CTA (2001), Chapter 1, pages 10–16.

was low and quality problems persisted related with crop management and the use of traditional tobacco varieties that were not the ones most accepted by consumers. Tobacco institutes were set up in each of the seven tobacco-producing areas, technical dissemination posts were established in each of the 50-odd tobacco-producing counties, and farmer groups were organized in order to provide technical assistance. Experienced tobacco farmers were recruited to create a technical assistance scheme composed of both agricultural technicians and experienced tobacco farmers.[15]

As a result of these efforts, the agroprocessing industry developed a set of production technologies appropriate to the conditions of Yunnan. All farmers used tobacco seeds provided by the agroprocessing industry and planted the quota according to the cigarette formula. In addition, the Yunnan Hongta Group provided technical assistance to farmers in all the phases of the tobacco cultivation. This resulted in a great increase in production, which doubled between the early and late 1990s, as well as an increase in quality, with tobacco leaves of high quality reaching 80% of total production.

An important issue that may have great effects in Yunnan's tobacco is the future trend in the demand for tobacco, in the context of public concerns about the harmful health effects of smoking and the implementation of anti-smoking policies both in industrialized and in many developing countries. In fact, jobs in the tobacco sector in the industrialized countries and in some developing countries have either been stagnating or declining, although tobacco and cigarette production have increased due to higher demand worldwide, as leading multinational tobacco companies capture new markets in both developing countries and countries in transition. These problems have even led to changes in the policies of international financial institutions. The World Bank has implemented a formal policy since 1991 that prohibits from lending on tobacco and encourages control efforts. The International Monetary Fund has also been imposing conditionalities on loans, in particular the privatization of tobacco state monopolies.[16]

4.3.2 Institutional Arrangements for Governing the Upgrading

The two cases studied here show that production and social upgrading took place based on different governance arrangements. These different arrangements were influenced by particular social characteristics of the two regions, government policies and programs.

Agroprocessing firms and contract farming in the Chinese case   The establishment of agroprocessing firms that promoted the cultivation of high-value crops through contract farming played the key role in the organization of tobacco production in southern Yunnan. Contracts were signed between the Hongta Tobacco

---

15   Maeda et al. (2003).
16   International Labour Organization (2003).

Group and individual farmers, establishing the commitment of the farmer to use the tobacco seeds provided by the firm, follow the technical directions provided by the firm's technicians during the tobacco cultivation and sell all the harvested tobacco leaves to the firm. Meanwhile, the firm committed to purchase the tobacco production at an agreed price and to provide training, technical assistance and credit in the form of inputs paid at the time of harvest and, in the case of tobacco, the construction of drying houses.

Contract farming played a key role in the development of tobacco because, in contrast to individual small farmers, the agroprocessing firm had better access to financial resources, know-how in agricultural technology and capacity to operate in changing product markets. In addition, the Hongta Tobacco Group had great access to local government and was able to obtain their support to construct and improve roads.

The key role of contract farming in the tobacco cluster in Yunnan relates mainly to the intensive use of labor characteristic of tobacco production and the high costs of monitoring wage labor that the tobacco industry would face if they grew tobacco themselves. Thus, tobacco firms in Yunnan preferred to produce under contract farming arrangements than to produce themselves. Farmers' compliance with the contracts (which is usually a concern for the industry because it may impose substantial costs) was high because of the strong relationship between the state-owned tobacco firms, local governments and community leaders in the tobacco-producing areas. Thus, farmers growing tobacco and not complying with the contracts would face negative consequences in their capacity to access other government services and benefits, especially those provided by local governments.

Several analysts have been highly critical of contract farming, arguing that it has frequently been characterized by low prices received by small farmers mainly due to their limited access to information and capacity to negotiate with firms.[17] However, it must be recognized that most if not all of the households working with cash crops that were visited in Yunnan said that they were better now than before cultivating them. The positive impacts of contract farming in Yunnan are in line with other studies, which have argued that contract farming and outgrower schemes have very often led to a significant rise in living standards.[18] In addition,

---

17   Several authors have considered contract farming in a dependency theory framework as an exploitative extension of international capital. For example, see Lappe and Collins (1977). Watts (1994) has considered contract farming as a system for self-exploitation of family labor and frequently characterized by company manipulation and abrogation of contracts. Little and Watts (1994) argue that the problems arising from unequal power relationships as well as market fluctuations, often make contract farming unsustainable in the long term.

18   For example, see Glover (1983 and 1987), Glover and Ghee (1992), and Glover and Kusterer (1990).

comparative experience shows that some of the problems of contract farming can be dealt with the creation of farmers' organizations that make collective bargaining and measures that strengthen their negotiation capacity possible.

Small farmer organizations in the Mexican case   In Chiapas, it was a farmer association (ISMAM) which played the key role in introducing the organic methods of production, in making possible for small farmers to cultivate organic coffee, and in promoting improvements in the quality of coffee. The role of the farmer association was important in the following ways:

a. *Organizing the marketing of organic coffee.* ISMAM played a key role in the marketing of coffee because foreign buyers preferred to negotiate with a farmer associations rather than with isolated individual farmers, an alternative that would have been too costly. In addition, buyers wanted to deal with a reliable organization that delivered organic products of good quality and was able to carry out the tasks required to ensure that the products were obtained according to organic standards. In fact, small-farmer associations have been dominant in both production and marketing of coffee in Mexico, with only a few firms producing and selling organic coffee. Associations of coffee producers are large, well organized, and highly experienced in dealing with foreign buyers, having learned how to deal with these buyers directly, as well as the importance of building a relationship based on trust and respect for agreements. Thus, they have been able to organize efficient marketing systems, and to establish quality standards to ensure that the exported product meets buyer expectations.

b. *Setting up monitoring systems to ensure compliance with organic standards of production.* The international norms that regulate the organic certification of products from groups of small farmers have established the obligation that the group must organize what is called an "internal control system", or monitoring system designed to collect and organize detailed information about the association and its members. The capacity of ISMAM to organize efficient and reliable monitoring systems was a key to success as organic producers. A well-functioning monitoring system must be able to identify easily and quickly those farmers who are not complying with the organic standards of production and penalize them in an exemplary way. A good monitoring system must make it very risky for an individual farmer not to comply, since he can be readily discovered and the losses he would suffer because he can no longer sell to the association would be devastating. In contrast, a monitoring system that does not work well and cannot adequately identify those farmers who are not complying with organic standards is likely to leave room for free riders who seek to obtain the better prices of organic products without paying the higher costs. ISMAM created its system between 1991 and 1994. The monitoring process has worked very well because ISMAM has trained all its members intensely about the potential negative effects of the use of chemical inputs and the fact that, if a single farmer does not comply with the organic standards, the association could lose its markets and the farmers would receive substantially lower prices. The interviews showed that, because the farmers had a good understanding of the potential negative consequences of free riders, most were willing to complain about any member who did not comply with the organic standards.

c. *Establishing connections with government agencies and NGOs.* ISMAM played an important role in the success of organic coffee by helping establish links with government agencies and NGOs, often in foreign countries. These links helped the association acquire important market information and access to funding and technical assistance. In fact, ISMAM determined soon after its creation that organic production might be an attractive alternative. It was able to do this because of its contacts with foreign buyers of coffee. Through a Catholic priest who had worked with ISMAM producers and later became the main person responsible for negotiating with potential buyers, ISMAM established contacts with European buyers and NGOs. Through these contacts, ISMAM learned of the opportunities in the organic market, which was expanding rapidly. Some of the buyers of conventional coffee were also interested in buying organic coffee, so, through them, ISMAM contacted organic certification agencies, which eventually certified ISMAM, in 1993. In addition, ISMAM obtained substantial support from government agencies and programs, which provided the funds ISMAM needed to purchase the coffee of its members, finance the production of its members and carry out investments to improve the processing capacity of the association.

The key role of a small-farmer organization in the upgrading of coffee production in Chiapas relates to particular existing social and production conditions (notably the dominant presence of an indigenous population that produced coffee with little or no use of chemical inputs), the promotion work of the Catholic Church in organizing a farmer association among producers of indigenous origin, government policies and programs that provided heavy support to farmer associations and the demand from buyers in industrialized countries for coffee produced under organic methods of production.

4.4 Creating Conditions for the Success of the Clusters:
Government Policies and Public Investments in Public Goods

In Yunnan, counties and municipalities in southern Yunnan played an active role in attracting new businesses. In recent years, the Chinese central government pushed local governments strongly to attract foreign and national investments. As a result, many county governments created Business Bureaus and provided benefits (like tax benefits and concessions for the use of public lands) to attract investors. In southern Yunnan, local governments focused on attracting investments of agricultural and agro-processing companies because they recognized that the characteristics of their counties made it very difficult to attract manufacturing industries, and at the same time that they had a great availability of natural resources. Part of the efforts focused on investments in upgrading or constructing roads connecting villages with small- and medium-sized towns and cities, as the commercial firms had told them that it was too difficult if not impossible for them to work with farmers in villages located in remote places or without good roads. Technicians from the tobacco company who were interviewed stressed that the availability of good roads was one of the important factors that they considered in the decision to incorporate a village to tobacco production. If the quality of roads was poor, rain could prevent the transportation of harvests to the factories, leading to product losses. In addition, local governments provided long-term concessions of public lands to produce cash crops like tea. More important, the county governments engaged actively in collaborative relations with the companies of tobacco, tea, and sugarcane that established in their counties, convincing farmers to introduce them, using village leaders as first growers to generate trust among other farmers, and training village leaders to help in the provision of technical assistance to other farmers.

One of the important factors that influenced the policies of local governments in Yunnan was that tobacco represented for them an important source of revenues. Unlike other agricultural products, tobacco and cigarettes are a government monopoly and heavily taxed. The two most important taxes on tobacco and cigarettes have been the "special crop tax" that tobacco companies pay to county government and the value-added tax paid when the company resells the tobacco directly to cigarette factories or to allocation stations. The special crop tax was imposed in 1983, accounting for 38 % of the gross purchasing price paid to farmers, and decreased to 31 % after 1993.[19]

In the case of organic coffee in Mexico, producers were not supported by government policies, programs, or agencies specifically working in organic agriculture. However, they benefited from the support of several government agencies and programs in agricultural and rural development in general, which made an important contribution to individual conventional farmers who undertook the shift to organic agriculture by providing grants and subsidized credit for

---

19   Special crop taxes were imposed on cash crops, in the context of insufficient food production. Thus, its purpose was to prevent the expansion of cash crops over lands dedicated to food production. According to Peng (1996), tobacco companies in China pay several other taxes, including among others taxes on net profits paid by the tobacco processing companies, VAT paid by allocation stations when they re-sell the tobacco to cigarette factories, VAT and consumption taxes paid by cigarette factories, and VAT paid by distributing companies when they sell the cigarettes to wholesale stations.

investments and inputs and technical assistance that helped the farmers introduce the new technologies. The most important sources of support included:

a. *Government agencies and programs supporting associations of organic producers.* This included the Alliance for the Countryside Programme and the National Fund for the Support of Solidarity Enterprises (FONAES), which was one of the most important interventions of the Mexican Government in support of agricultural and rural development during the nineties, being funded jointly by the federal and state governments. The Alliance for the Countryside Programme provided subsidies to individual farmers and farmer associations for investments, inputs, technical assistance, training and research in a great variety of agricultural activities. In coffee, support included the replacement of old plants with new improved varieties, the implementation of phytosanitary practices, the purchase of equipment by individual farmers and farmer associations and the contracting of extension services. During 2000 and 2001, through the Programme for the Support of Coffee Production, 262 ISMAM members received help in the production of seeds and plantings, 10 in the construction of 10 drying yards, 21 in the purchase of pulp extractors and others in the renovation of coffee plantations. Meanwhile, the National Fund for the Support of Solidarity Enterprises (FONAES) aimed at strengthening farmer associations through subsidized credit (zero nominal interest rates), being implemented since 1992 by the Secretariat of Solidarity Development, being transferred in the late 1990s to the Secretariat of the Economy. FONAES focused on three areas: (i) the provision of credit for labor and inputs to informal farmer groups lacking access to formal credit; (ii) credit for formal farmer associations that is usually employed to build marketing and processing facilities and to purchase product from members and (iii) grant funds to contract technical assistance. ISMAM received funds through all three focal areas of FONAES since 1992, which it used to purchase storage and processing facilities, modernize processing facilities, supply credit to farmers for the maintenance of coffee plantations, expansion of areas under coffee cultivation, purchase of inputs and wages.

b. *Rural Credit.* One of the problems that organic producers often face is the difficult access to credit from formal financial institutions. If they have little experience in organic methods of production, producers of organic crops face problems similar to those of producers of new products, who have difficulty getting credit because financial institutions find it risky to finance a business which has little experience and knowledge. Banrural is a public bank that has had an important role in financing in the agricultural sector. According to the Banrural manager in Tapachula, the share of organic coffee in the credit provided to coffee growers over the last few years of the 1990s averaged 60% of the total in the Tapachula branch and 90% in the Motozintla branch. Banrural even determined the different production costs of organic and conventional coffee, so that it could take into account these differences. The role of Banrural was relevant because it provided credit both to ISMAM and to individual ISMAM members since the initial stages in the life of the association.

4.5 Conclusions and Policy Implications

On Global Standards and Social Upgrading

The evidence from the cases presented here shows not just that global standards play a key role in the social impacts of the development of agriculture-based clusters, but that these effects can have different directions. The case of organic coffee in Chiapas suggests that the increasing international demand for organic products had positive impacts on the coffee cluster, as the organic market made it possible for small farmers to obtain better prices for their product, while producing with more environmentally-friendly technologies. In the case of tobacco in Yunnan, the expansion of the crop in the southern portion of the province created a new source of income among small farmers. However, public concerns about the negative health effects of smoking and the implementation of anti-smoking policies elsewhere generate questions regarding the future trend in the demand for tobacco. Thus, farmers may need to find new profitable crops in the medium to long term.

On Contract-Farming Schemes

Contract farming involving processing and marketing firms and small farmers have facilitated the marketing of small farmers output and given farmers access to extension services and occasionally to credit, but they have also had disadvantages. Small farmers have a relatively weak position in negotiations with firms because they have limited information and are poorly organized, so they have ended up receiving relatively low prices and accepted contract terms that have not been convenient for them. In addition, outgrower schemes with small farmers may have severe limitations, including the high costs of monitoring the contracts with small farmers and the difficulties in appropriating the benefits of the investment in the schemes due to the diversion of output to other buyers who may be paying higher prices than the ones agreed in the contracts. Government programs or projects may support the organization of small farmers, as well as provide legal and managerial support to them in order to strengthen their capacity to negotiate contracts with agroprocessing firms. In addition, the case of tobacco in Yunnan shows that local government played an instrumental role in the development of the cluster, by attracting the firms to establish in their counties, by providing investments in rural road infrastructure that were essential for the transportation of the product to the industry, and by facilitating the negotiations between firms and small farmers.

On Small-Farmer Associations

The successful case of organic coffee production in Chiapas shows that producer associations played a key role in the social positive effects associated with the development of the cluster. This role relates to the following: a) the generation of advantage of economies of scale in the collective marketing of the products of their members and in managing volumes that attract foreign buyers to negotiate with them; b) their capacity to train a large number of small farmers in the basics of organic production and to promote among the farmers the adoption of new production technologies required by the organic production standards; c) the implementation of monitoring systems to control the compliance of their members with the standards of organic production. This evidence suggests that the development of small-farmer organizations focused on the marketing of production, the dissemination of technologies among their members, and the monitoring of the compliance of members with the global standards of production, can play an important role in generating a socially inclusive model of the agricultural cluster. It must be recognized that there is a long history of farmer organizations (cooperatives, associations, etc.) that have not worked well. Thus, their support should target the specific constraints that they face, focusing specially on the strengthening of managerial and organizational skills.

References

Asian Development Bank (2005). *Pathways Out of Rural Poverty and the Effectiveness of Poverty Targeting.* Special Evaluation Study. Manila: Asian Development Bank, Operations Evaluation Department.

Damiani, Octavio (2001). Organic Agriculture in Mexico. Case Study of Small Farmer Associations in Chiapas and the Yucatan Peninsula. Rome: International Fund for Agricultural Development.

Damiani, Octavio (2002). *Small Farmers and Organic Agriculture. Lessons From Latin America and the Caribbean.* Rome: International Fund for Agricultural Development.

Damiani, Octavio (2006). Rural Development from a Territorial Perspective: Case Studies in Asia and Latin America. Background paper for the World Development Report 2008. Washington, DC.

FAO/ITC/CTA (2001). *World Markets for Organic Fruit and Vegetables: Opportunities for Developing Countries in the Production and Export of Organic Agricultural Products.* Rome: FAO/ITC/CTA.

Glover, D. (1987). Increasing the benefits to smallholders from contract farming: Problems for farmers organisations and policy makers. World Development, 15 (4), 441–48.

Glover, D. (1983). Contract farming and smallholder outgrower schemes in less developed countries. *World Development,* 12 (111–12), 1143–57.

Glover, D. and Ghee, Lim Teck (1992). *Contract Farming in Southeast Asia: Three Country Case Studies*, Kuala Lumpur: University of Malaya.

Glover, D. and Kusterer, K. (1990). Small Farmers, Big Business: Contract Farming and Rural Development. London: Macmillan.

International Labour Organization (2003). Employment Trends in the Tobacco Sector: Challenges and Prospects. Report for discussion at the Tripartite Meeting on the Future of Employment in the Tobacco Sector. Geneva.

Lappe, F. M. and Collins, J. (1977). *Food First: Beyond the Myth of Scarcity*. San Francisco, CA: Institute for Food and Development Policy.

Little, P. and Watts, M. (1994). *Living Under Contract: Contract Farming and Agrarian Transformation in Sub-Saharan Africa*. Madison, USA: University of Wisconsin Press.

Maeda, Yuko; He, Yongqing and Zhang, Yunling (2003). A study of the tobacco sector in selected provinces of Cambodia and China. Working Paper No. 185. Geneva: International Labour Office.

Peng, Yali (1996). The Politics of Tobacco: Relations Between Farmers and Local Governments in China's Southwest. *The China Journal,* 36, 67–82. Contemporary China Centre. Canberra: Australian National University.

Watts, Michael J. (1994). Life under Contract: Contract Farming, Agrarian Restructuring, and Flexible Accumulation. In Peter D. Little and Michael J. Watts (eds), *Living under Contract: Contract Farming and Agrarian Transformation in Sub-Saharan Africa*, 21–77. Madison, USA: University of Wisconsin Press.

Chapter 5

Small Firms in the Indian Software Clusters: Building Global Competitiveness

Aya Okada[1]

5.1 Introduction[2]

In the 1990s, software clusters emerged in several regions of India. These clusters, most notably Bangalore, have recently grown remarkably as a major destination of global outsourcing for software development and services. The Indian software industry depends on exports for about 80% of its income, making it India's leading export industry. Initially, the industry achieved its impressive growth, by exporting relatively low value-added software services to foreign markets, especially the US, taking advantage of the abundant supply of relatively low-wage and English-fluent IT workers. In more recent years, however, many global IT firms have started operations in various software clusters in India, leading to a considerable shift in focus to higher value-added activities such as R&D, and rapidly expanding activities in IT-enabled services (ITES) and business process outsourcing (BPO).

1   Aya Okada is professor of political economy, at the Graduate School of International Development, Nagoya University, Japan. Her academic training was in regional planning and international economic development. Her current research interests include regional industrial development, regional economic development policy, and training and skills development in Asia. Dr Okada holds a M.Phil. from the University of Sussex, the UK, and a Ph.D. from Massachusetts Institute of Technology (MIT), the USA. e-mail:aokada@gsid.nagoya-u.ac.jp

2   Acknowledgement: This research was conducted with financial support from the Japan Society for the Promotion of Science (JSPS) under its grants-in-aid (Category (C)(2): No. 16530151) for 2004–2007 and a research fellowship awarded in 2004 from the Ministry of Finance of the Japanese Government. An earlier version of this chapter was presented at the International Workshop on "Upgrading Clusters: Experiences of Asia and Latin America," Rio de Janeiro, Brazil, August 24–25, 2006; the International Seminar on Knowledge-based Industries, Employment and Global Competitiveness, at Delhi, India, October 6–7, 2006; and the 48th Association of Collegiate Schools of Planning (ACSP) Annual Conference, Milwaukee, October 18–21, 2007. I thank participants in these conferences for their helpful comments. I also thank many people from the software industry, government agencies, and industrial associations in India who generously shared with me their knowledge, experience and insights. An earlier version of this chapter was first published as Chapter 3 of High-tech Industries, Employment and Global Competitiveness, edited by S.R. Hashim and N.S. Siddharthan, 2008, London, New Delhi: Routledge, pp. 43–69. I thank Routledge for its permission to reprint this chapter.

While a handful of large domestic firms and multinational corporations (MNCs) played a key role in driving the growth of both the industry and its exports, many small start-up firms were established in these clusters. These newly emerging small firms, which have better technological capabilities and focus more on exports, differ considerably from the traditional firms in other sectors, which focused on domestic markets and were heavily protected under the decades-long Indian government policies on small-scale industries (SSI).

Given this context, what challenges do these newly-established small firms face to compete in the global market? How do these small firms develop their capabilities to innovate and build their competitiveness in both the domestic and global markets? What linkages do they form with other firms, including global firms, within and outside the clusters? If a small number of domestic large firms and MNCs are so dominant, what roles do domestic small firms play in the global division of labor? This chapter addresses these questions.

In knowledge-based industries such as software, as the sources of competitiveness largely depend on the creation, transmission and sharing of knowledge, it becomes critically important to build mechanisms that can be used to create, transmit and share that knowledge. Thus, if small domestic firms are to promote innovation, become more competitive, and shift to higher value-added activities in the global division of labor, it is important to build such internal mechanisms within the firm as well as within the cluster for skills and technological upgrading.

The objective of this study is thus twofold. First, the study analyzes the internal structure of the Indian software clusters and investigates the relative importance of domestic firms in promoting the development of regional clusters. Second, it examines the patterns of capability development for building competitiveness among small firms within the cluster.

To examine these questions, I use a case study of the Indian software industry, especially in Bangalore in the state of Karnataka, the oldest and largest software cluster in India. This chapter draws on extensive interviews with about 40 software firms of different sizes in Bangalore, Delhi and NOIDA (in Uttar Pradesh) that I carried out during three rounds of fieldwork in India between 2001 and early 2005. During these visits, I also interviewed with and collected information from central government institutions as well as those in the state government of Karnataka, and industrial organizations such as the National Association of Software and Service Companies (NASSCOM) and the Electronics and Computer Software Export Promotion Council (ESC).

This chapter is organized as follows. Section 2, drawing on the existing literature, conceptually discusses the sources and nature of competitiveness of firms clustered in knowledge-based industries, and the challenges that small firms face in building this competitiveness. Section 3 identifies the patterns of geographical distribution of the software firms and clusters in the Indian software industry. Section 4 analyzes the strategies that domestic small firms in software clusters have adopted to build their competitiveness, focusing on the patterns of their capability development, and of inter-firm linkages between small domestic firms and their affiliates and clients both within India and abroad. Section 5 summarizes the findings and discusses their implications for policy.

5.2 Building the Competitiveness of Firms in Knowledge-based Clusters

Since the 1990s, with globalization and technological change that have occurred at an unprecedented speed, and with increased importance of technological progress embodied in manufactured goods as well as in services, competition has intensified in the global market. Thus, competitiveness has become an important policy concern for both developed and developing countries. Meanwhile, the key determinants of competitiveness are also changing rapidly. As the global economy increasingly becomes knowledge-based, the economic performance of countries and regions depends on a series of relatively immobile resources such as knowledge, technologies and skills, and institutional and organizational structures (Breschi and Malerba, 2001: 817). The development of information and communication technologies (ICT), and the diffusion of the Internet in particular, have made it easier to network, collaborate and share knowledge among firms across national boundaries. This has reduced the costs of transportation, communication, affiliations with foreign firms and outsourcing from overseas clients.

In turn, a new pattern of competition has emerged, based on knowledge and technological advantages, rather than on traditional competitive advantages based on inherent factor endowments. Innovation has become more and more important as a determinant of competitiveness, for developed and developing countries alike. Indeed, the Organization for Economic Cooperation and Development (OECD) observes that "innovation has become more market-driven, more rapid and intense, more closely linked to scientific research, more widely spread throughout the economy" (OECD, 2000: 8).

In order to build competitiveness, therefore, firms, and industries, need to develop their capabilities to innovate. To do so, they must accumulate a set of adequate knowledge and skills that allows them to innovate, and to develop markets for goods and services as well as for finance. Thus, for developing countries to shift towards knowledge-based economies, and for developing-country firms to promote innovation, active public policies may be necessary to ensure the conditions conducive to innovation. In particular, in knowledge-intensive industries such as software, human resources are the main inputs; and thus the knowledge and skills embodied in these human resources are an important source of innovation. For all these reasons, it is important to generate, transfer, and diffuse adequate knowledge and skills among firms, and to upgrade them at national, regional (cluster), and firm levels. Porter (1990) argues that continuous improvement and technological upgrading are the most important conditions to keep firms competitive. Thus, building firms' competitiveness must necessarily involve the process of technological learning within and between the firms.

Indeed, challenging the conventional neoclassical thinking that undermines the processes of becoming efficient and of learning to use technologies, evolutionary theorists argue that the vital process of learning to become efficient entails considerable skills development and institutional arrangements (Nelson and Winter, 1982). They also stress that it is not just formal R&D that brings firms to promote innovation, but informal activity at every level in a firm (Nelson and Winter, 1982).

Firms can promote innovation by finding new interpretations for existing information, by adopting new ideas that modify the existing knowledge, and by bringing new people from other industries to the team (Porter, 1990). Moreover, firms do not learn on their own; they often draw on other firms for new information, knowledge, and skills. Especially in developing countries, firms often bring such new interpretations and ideas from abroad. Foreign direct investment (FDI) is clearly an important channel for doing so. Also, experiences in developing countries suggest that start-up firms are important generators of new ideas and innovation, and may have an advantage over large established firms in emerging areas where demand is uncertain and risks are high (OECD, 2000: 9).

Thus, it is important to develop a "forum" for firms to create, share, transmit and adopt new sets of information, insights, interpretations and ideas, and develop global linkages that would entail inter-firm cooperation. Some argue that the creation of such a forum, or what some scholars call the "regional innovation system (RIS)" (Cooke, 2001; Breschi and Malerba, 2001), involving linkages, networking, and mutual learning among various local organizations such as firms, universities, research institutes, science parks and technological transfer centers, is a key factor contributing to the development of industrial clusters.

It is well-known that knowledge spillovers and externalities play very important roles in the growth of knowledge-intensive industrial clusters (Breschi and Malerba, 2001, 2005). In order to promote them, it is critically important to develop the internal mechanisms that will let firms and other organizations create, share, transmit and adopt information and knowledge. Industries with international competitiveness facilitate the development of suppliers and other supporting industries by transferring skills and spillovers (Porter, 1990). As firms from other industries enter a cluster, they are all more likely to diversify their R&D, develop new strategies, form new sets of skills and share information; in turn they create new ideas and upgrade skills, all of which further induces innovation (Porter, 1990).

In the literature, earlier constructs of industrial clusters, or industrial districts, focused on dynamic, innovative small firms, which typically forged cooperative horizontal networks among themselves, and played a considerable role in promoting regional growth (Piore and Sabel, 1984; Pyke et al., 1990). Others identified alternative models of industrial clusters, which depict the dominance of the state and/or MNCs in shaping and leading these clusters (Markusen, 1996). In these alternative models, MNCs provide "the glue that makes it difficult for smaller firms to leave, encouraging them to stay and expand, and attracting newcomers into the region" (Markusen, 1996: 294). We might see this phenomenon in industrial clusters in many developing countries, which are currently absorbing growing amounts of FDI.

However, few studies have focused on the role of small- and medium-sized enterprises (SMEs) within FDI-driven knowledge-intensive clusters in developing countries. SMEs tend to face more constraints than large firms, especially in gaining access to the global markets, responding to the market demands and providing services. But, in developing-country clusters, SMEs may face even greater constraints as they try to develop overseas channels, develop innovative capabilities, and acquire the right set of skills. But so far we know little about how SMEs operate in knowledge-intensive industrial clusters in developing countries: what measures do they actually adopt to build their competitiveness?

5.3 The Structure of the Indian Software Clusters and the Patterns of Agglomeration

5.3.1 Software Clusters in India

In addition to well-known Bangalore, which is the largest and oldest software cluster in the country, several other software clusters have emerged in various parts of the country. These clusters include: National Capital Region (NCR, which includes Delhi, Gurgaon in Haryana, and NOIDA in Uttar Pradesh); Hyderabad (Andhra Pradesh); Ahmedabad and Gandhinagar (Gujarat); Kochi and Thiruvananthapuram (Kerala); Mumbai, Navi-Mumbai and Pune (Maharashtra); Chandigarh (Punjab); Jaipur (Rajasthan); Chennai (Tamil Nadu); and Kolkota (West Bengal).

The Indian software industry estimated that over 6,000 software firms exist in various locations; of these, about 4,400 are members of the Software Technology Park of India (STPI).[3] Table 5.1 shows the size distribution of software firms in India as a whole in 2000/01. The table clearly shows the pyramidal structure of the Indian software industry. The majority of firms are small in size, earning only less than Rs 100 million (approx. US$2.2 million) as annual turnover. In recent years, only this category of smaller firms grew in number, while the number of larger firms—those earning more than Rs 1 billion (approx. US$22 million)—remained largely unchanged.

---

3   In 1991, the Indian government set up the Software Technology Park of India (STPI) in 29 cities, as an autonomous body under the central government. STPI provides new start-up firms that are registered with STPI with a range of services such as infrastructure and communications on a priority basis, tax breaks for the first five years, and provision of new technologies. In return, these firms are mandated to export their products. For detailed discussions on the role of STPI in the growth of the Indian software industry, see Heeks (1996), Parthasarathy (2000), and Saxenian (2001). At present, STPI is established in 35 cities and the central government plans to set it up in every state.

Table 5.1   Structure of the Indian software exports industry

Annual Turnover (Rs in million)	No. of Firms (2000/01)	No. of Firms (2001/02)
Above 10,000	5	5
5,000–10,000	7	5
2,500–5,000	14	15
1,000–2,500	28	27
500–1,000	25	55
100–500	193	220
Below 100	544	2,483

Note: Rs 48 = approximately US$1 for the year concerned.

Source: NASSCOM 2003.

But, this small number of large firms is playing a leading role in promoting software exports, as well as in developing software clusters in different regions, as they typically have operations in multiple locations. The top five firms have sales of over Rs 10 billion (approx. US$220 million) each; interestingly, these are all Indian firms, rather than MNCs. They account for nearly a third of the total software exports from India, and 45% of largely low-tech IT-enabled services (ITES) (Sridharan, 2004). Thus, interestingly, the same firms often undertake both high-end services involving sophisticated systems integration and low-end ITES, such as call centers. While many Indian firms achieved their growth based on their global or offshore service delivery model, these top firms and a few others, like Ramco and Iflex, are moving up the value chain by developing proprietary software products and offering high-end consulting services (CMIE, 2004: 3).

In addition, some firms are captive ITES/BPO operations that work only for their affiliates abroad; these include GE Capital International Services, Standard Chartered Bank's Scope International, and American Express. Moreover, many MNCs have operations in India that exclusively service their own plants in other countries, using them as cost centers. Indeed, many global IT firms, especially MNCs, have accelerated the outsourcing of their R&D operations to India in recent years. Leading IT firms such as IBM, Intel, Sun Microsystems, and Texas Instruments (TI) have all located some of their operations, especially their R&D, in India.

With regard to the geographic location of these software firms, interestingly, the larger firms have multiple operations, with offices in various software clusters, while the smaller ones mostly undertake single or multiple operations within a cluster. For example, Tata Consultancy Services (TCS), the largest IT firm in India, has offices in seven main software clusters: Bangalore, Delhi, Mumbai, Pune, Kolkota, Chennai and Hyderabad. Each specializes in a different line of activities.[4] Likewise, Infosys, the second largest firm, has offices in 11 locations; four are in Karnataka, including its head office in Bangalore. Similarly, Wipro Technologies, the third largest, has operations in five clusters within India, including Bangalore. Thus, the patterns of these firms' locations clearly differ from those of manufacturing firms such as automobile assemblers,[5] which tend to locate their plants in a single or at most in a few regions. By contrast, these large software firms are present in every major software cluster across the country.

As the US software market has expanded, the Indian software and services industry has grown remarkably, with total export earnings reaching US$ 175 million in 1989/90, to US$ 5.7 billion, and further to US$10.01 billion in 2002/03 (NASSCOM, 2004). Table 5.2 shows the size and recent export performance of these clusters: overall, India's software clusters have experienced rapid growth in exports in recent years, especially in the last few years. As the Indian software industry caters mainly to the export market, and virtually all exporting firms are members of STPI, the figures for STPI members presented in Table 5.2 provide a fairly accurate picture of the economic performance of major software clusters in terms of export earnings for the last five years. In recent years, the export market for Indian software clusters has grown much more rapidly than the domestic market.

Table 5.2 clearly shows that a rising number of firms are located in newly emerging software clusters like Hyderabad and NOIDA, but Bangalore has kept its dominant position as the largest software cluster in terms of the exports. Its export earnings reached Rs. 181 billion (US $3.9 billion) in 2003/04, accounting for nearly one third of India's total software exports. It also suggests that the spatial concentration of economic activities, as measured by export performance, has intensified in selected clusters such as Bangalore and Hyderabad.

However, more than 50% of software and services exports are in the form of customized software application. On the other hand, more knowledge-intensive work, such as turnkey projects and IT consulting, accounts for only 3% and 19% respectively. Moreover, few products have a global brand (Government of India, 2001), as Indian software products account for only 0.2% of the global software products market (NASSCOM, 2002).

As the Indian software industry is gradually moving up the value chain, subsectors with higher value-added—such as embedded software/ technologies, chip design and telecommunication—are growing. Moreover, not only IT firms but also global manufacturing firms such as those producing electrical goods and auto components, are increasingly moving to India to outsource their R&D operations. The main destination of the Indian software and services exports has been the US, accounting for 60% of total exports, followed by the UK, which accounts for 13.5%. Since 2001, when the US economy stagnated after the IT

---

4   TCS internal presentation material, December 2004.

5   For detailed discussions of the geographical distribution of the Indian automobile firms, see Humphrey et al. (1998) and Okada (2004).

Table 5.2   Software clusters in India: Distribution of exporting firms and recent export performance (in Rs million)

Software Clusters	# of Exporting Firms* (as of Dec. 2004)	Export Earnings in 1999/00	Export Earnings in 2000/01	Export Earnings in 2001/02	Export Earnings in 2002/03	Export Earnings in 2003/04	Growth Rate 2002/03 to 2003/04
Bangalore	705	43.2 (25.0%)	74.8 (27.2%)	99.0 (27.1%)	123.5 (26.6%)	181.0 (31.2%)	47%
Bhubaneshwar	46	0.9 (0.5%)	2.0 (0.7%)	2.1 (0.6%)	2.6 (0.6%)	3.2 (0.6%)	23%
Chennai	535	18.9 (10.9%)	29.6 (10.7%)	50.2 (13.8%)	63.2 (13.6%)	76.4 (13.2%)	21%
Gandhinagar	89	0.3 (0.2%)	1.0 (0.4%)	1.2 (0.3%)	1.1 (0.2%)	1.4 (0.2%)	34%
Hyderabad	785	10.6 (6.1%)	19.9 (7.2%)	28.1 (7.7%)	36.7 (7.9%)	50.3 (8.7%)	37%
Kolkota	103	1.5 (0.9%)	2.5 (0.9%)	6.0 (1.7%)	12.0 (2.6%)	16.0 (2.88%)	33%
Mumbai	445	9.6 (5.6%)	16.1 (5.9%)	26.0 (7.1%)	27.1 (5.8%)	43.2 (7.4%)	59%
NOIDA	782	24.8 (14.3%)	44.2 (16.1%)	60.9 (16.7%)	76.0 (16.3%)	99.0 (17.1%)	30%
Pune	306	5.7 (3.3%)	9.6 (3.5%)	20.0 (5.5%)	28.0 (6.0%)	42.0 (7.2%)	50%
Thiruvananthapuram	114	0.6 (0.3%)	0.9 (0.3%)	1.6 (0.4%)	1.7 (0.4%)	2.1 (0.4%)	28%
Sub-total (STPI members in the clusters above)	3,910	116.7 (67.1%)	200.5 (72.9%)	295.2 (80.9%)	371.7 (79.9%)	514.6 (88.7%)	38%
STPI members in other locations and non-STPI members	n/a	56.9 (32.9%)	74.5 (27.1%)	69.8 (19.1%)	93.2 (20.1%)	65.4 (11.3%)	-30%
All India	n/a	173.0 (100%)	275.0 (100%)	365.0 (100%)	465.0 (100%)	580.0 (100%)	20%

Notes: As the figures are rounded, the sum does not necessarily match the figures for the subtotal and the total. * The number of exporing firms in each location refers to STPI members in that cluster. Exports are mandatory for STPI member firms.

Source: The author's calculation using data from STPI internal documents and ESC (2005).

bubble burst, India has tried to diversify its export markets, focusing increasingly on Asian countries, and Japan in particular. Still, Japan accounted for only 4.4% of its export markets in 2004 (ESC, 2005).

5.3.2 The Structure of Bangalore's Software Cluster

Bangalore, the state capital of Karnataka, is the country's oldest and largest software cluster.[6] As of 2004, more than 1,322 software firms were located there.[7] As of 2003/04, Bangalore housed 110 MNCs, many based in the US and Europe, including major global players (STPI, 2004). The *Karnataka IT Directory 2001/02* listed 951 software and services firms in Bangalore; more than 60% were domestic SMEs (see Table 5.3). Although this total number of firms does not match with the STPI figure, it still helps us grasp the distribution of firms in the Bangalore cluster by firm size.

Table 5.3   **Distribution of software firms in Bangalore by size and ownership**

Size	Ownership	Number of Firms (%)		Subtotal
Large	MNC	217	22.8	253
	Indian	3	3.8	
SME	MNC	26	2.7	648
	Indian	622	65.4	
Not known		50	5.3	50
Total		951	100	951

Source: Computed by Author from *Karnataka IT Directory 2001/02*.

MNCs have become even more dominant in Bangalore in the last few years, with a rapidly growing move among many US firms (and some European firms) to outsource their R&D, as well as their ITES and BPO to Bangalore. MNCs had invested Rs 197 billion (US$ 4.1 billion) in Bangalore as of 2003/04, while Indian firms of all sizes had invested Rs 4.37 billion (US$ 92 million) (STPI, 2004). Despite a rapid growth in new start-ups, particularly among SMEs, the top 10 firms account for 50% of exports from Bangalore, whereas SMEs account for only 20%. Of Bangalore's firms, 35% have close ties with those in foreign locations

---

6   For the historical evolution of Bangalore's software cluster and the detailed analysis of factors contributing to the development of the cluster, see Heeks (1996), Parthasarathy (2000) and Okada (2005).

7   The Software Technology Park of India (STPI) – Bangalore (2004). Figure represents the number of registered firms under STPI, Bangalore. Thus, the actual total number of software firms should be larger than this figure.

through non-resident Indians (NRI).[8] Clearly, apart from the presence of leading Indian software firms such as Infosys and Wipro, Bangalore's software cluster has also been driven by its close connections with the global market through MNCs and NRIs.

Figure 5.1 shows the rapid export growth of Bangalore's cluster in the software and services industry between 1991/92 and 2003/04. With the remarkable growth of the software and services industry, the number of firms in the industry in Bangalore (and neighboring cities) increased from 13 in 1991/92 to 728 in 1999/00, and further to 1,322 in 2003/04. The value of exports increased from US $1.3 million to US $3.2 billion during this period, accounting for 36% of India's exports in this industry (STPI, 2004). Customers from the US account for 54% of Bangalore's software and services exports, while Europe and Japan account for 24% and 4% respectively. Clearly, as in Hsinchu in Taiwan (Saxenian and Hsu, 2001), these close links with the US market have contributed to the cluster's rapid growth.

Especially in the last few years, Bangalore's ITES/BPO segment, which includes banking and financial services, call centers, technical support and insurance claims processing, has grown rapidly, with the growth rate of 135% in 2003/04 over the previous year, generating more than 60,000 jobs in Karnataka. Indeed, out of 168 firms that started up newly in Bangalore in 2003/04, 26% were in ITES/BPO services, while 49% were in enterprise application services (STPI, 2004).

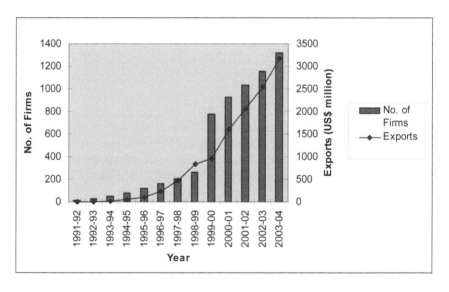

Figure 5.1   Bangalore: Software export growth

8   Interview with Joint Manager, STPI, Bangalore, in December 2001.

On the other hand, in recent years, MNCs have rapidly increased their R&D outsourcing. For example, Sun Microsystems and IBM created R&D centers in several locations in India, including Bangalore; and Motorola shifted 90% of the R&D functions for its mobile phone operation to Bangalore. Similarly, Hewlett Packard India (HPI) has a large R&D operation as part of its International Software Operation Department, mainly developing embedded software for HP's hardware products. Accordingly, Bangalore firms have shifted their interest toward high-end services such as embedded systems, design and manufacturing chips, and embedded communications (Okada, 2005). Thus, while the development of enterprise software applications continues to dominate, activities have diversified in Bangalore, requiring a wide range of skills.

Table 5.4 shows the distribution of software and services exports from Karnataka by segment. While ITES/ BPO currently accounts for only 13% of Karnataka's exports in the software and services industry, the share is likely to increase further in the coming years. However, despite the rapid recent growth in R&D services, less than 1% of all software firms are engaged in R&D services for exports, and the value of R&D services in the total sales is only 2% to 7 % (Sridharan, 2004). Although large firms exported almost US$1 billion in R&D services exports (Sridharan, 2004), these services are carried out only by a small number of leading Indian firms such as Infosys, Wipro Technologies, and HCL.

Table 5.4   Software exports by segment: Karnataka (2003/04)

Segment	Distribution (%)
Enterprise Application Software	36
Telecom	16
System Software	14
IT Enabled Services (ITES)/ BPO	13
Embedded Software	10
IC Design	9
Others	2
Total	100

Source: STPI (2004).

5.4 Bangalore's Small Firms: Building Competitiveness

In this section, I analyze the role of small firms in developing Bangalore's cluster and the patterns in their process of building competitiveness. Based on the discussion in Section 2, domestic SMEs' firm-level capabilities for building competitiveness are considered. Before doing so, however, I briefly discuss the division of labor within Bangalore's cluster.

5.4.1 The Division of Labor within the Cluster

The Indian software industry incorporates three general business models: a) turnkey projects; b) medium- to long-term contracts for largely offshore-based fixed-term consulting services; and c) body-shopping contracts based on time and labor. The Indian software industry has rapidly grown since the late 1980s, mainly undertaking tasks such as coding and programming, mostly for US clients. In the early 1990s, firms used the body-shopping business model, but since the mid-1990s, they have gradually shifted to the offshore outsourcing model, where most work is carried out in India (Parthasarathy, 2000; Patibandla and Pertersen, 2002). Indian firms still perform a small proportion of turnkey projects, as those require a higher level of integration and higher levels of both technical and managerial skills. Many large firms combine fixed-term consulting services and body shopping. Recently, however, they have been increasingly shifting their focus from body shopping to offshore-based fixed-term consulting services.

What activities, then, are SMEs in Bangalore's software clusters engaged in? As I discussed elsewhere, we can see a skill-based hierarchical structure among different lines of activities within the industry (Okada, 2005). However, firm size does not determine the type of activities the firm undertakes. Indeed, leading large firms such as TCS and Infosys engage in all levels of activities, ranging from low-end BPO to high-end R&D, although the largest chunk of their work is still in the category of customized software application. Likewise, small firms engage in activities in all the tiers. Unlike the large firms, however, each small firm tends to specialize in a relatively narrow range of activities. While some small firms are engaged in highly specialized high-end activities such as embedded software and IC chip design, including highly-specialized niche activities, such as VSI design, many others are engaged in relatively simple tasks such as application, data conversion and coding. However, small firms spend far less on R&D than large firms. Also, very small firms rarely engage in BPO activities like call centers which require a relatively large initial capital investment.

On the other hand, global IT firms like IBM and HP carry out a wide range of activities, from high-end R&D to low-end software application services, as well as to ITES/BPO. Other global IT firms like Oracle, Microsoft, SAP, and Adobe largely use their R&D functions for intra-firm exports of software services to their operations elsewhere. Moreover, by 2003, about 100 MNCs had established R&D centers in India as their 100% subsidiaries. Firms like TI and Intel have set up R&D centers to design IC chips in Bangalore, as the largest R&D center outside the US. Including these, a total of 230 MNCs have established R&D service centers in Bangalore, employing 25,000 Indian engineers, mainly developing telecommunication applications and chip design (Kriparalani, 2004, as cited in Sridharan, 2004).

The process of software development typically involves the following flow of value chains: 1) conceptualization; 2) requirement analysis and decisions on specification; 3) high-level integrated design; 4) low-level design; 5) coding and programming; 6) prototyping; 7) testing; 8) delivery, installation, and operationalization; 9) module test; and 10) customer support and maintenance. In the upper stream of the value chains are higher value-added activities. Since the late 1990s, more firms have been trying to move up the value chain by starting such activities as requirement analysis and design.

Within Bangalore's software cluster, a division of labor occurs among the firms engaged in various steps in the above-mentioned flow. MNCs located in Bangalore, such as Intel, TI, and Motorola, subcontract low-end R&D, including requirement analysis, applied research, and design, as well as relatively low-end activities such as programming and coding to local small firms, developing vertical supplier relations (Patibandla and Peterson, 2002; Sridharan, 2004). For instance, MNCs such as Iflex and Motorola have developed long-term, stable contractual relationships with local software firms, including many SMEs. Some MNCs have selected designated subcontractors. For example, Intel subcontracts about 60% of its software development to 20 to 25 local SMEs, although it carries out the entire IC chip design 100% by itself.[9] For high-end core R&D activities, however, MNCs carry out the work internally by themselves or send it to their R&D centers elsewhere. Similarly, Microsoft subcontracts about 70% of work to local SMEs.[10]

These global IT firms offer various opportunities to transfer the latest knowledge and technologies to their local suppliers and users. For example, Microsoft organizes seminars and workshops for local software firms in Bangalore that use the Windows platform, even before new Microsoft versions or products are released in the market, and it provides technical support for problem solving. Moreover, Microsoft has developed "key skill networks," so as to select the right local firms with the right skill sets when subcontracting its work. Likewise, Intel has set up its "Early Access Program." Through this program, the firm provides local software firms in contractual relations, including its suppliers, with new products for trial before they come out in the market. This helps Intel quickly rectify any problems that the local firms identify, before the products go to the market, and local firms also benefit from these trials by learning new features, knowledge, and technologies. Thus, the program brings benefits to both Intel and its local subcontractors, in terms of transferring and sharing information and knowledge, as well as reducing costs.[11]

The existence of these vertical inter-firm linkages suggests that Bangalore's pattern of agglomeration differs from the Marshallian model of industrial clusters, as in Sillicon Valley (Saxenian, 1994) and Emilia-Romagna (Piore and Sabel, 1984). Rather, it exhibits a pattern similar to that of manufacturing clusters such as automobiles (Okada, 2004). Interestingly, however, domestic large firms like Infosys and Wipro Technologies do not use local small firms as suppliers, because it is more profitable to expand their own operations internally than to subcontract

---

9   Interview at a software firm in Bangalore (November 29, 2004, Bangalore).
10   Interviews at a software firm in Bangalore (November 29, 2004, Bangalore).
11   Interview at a domestic software firm (December 2004, Bangalore).

some processes or some projects.[12] When these firms experience an increase in demand, they respond by hiring more employees, rather than subcontracting out the work that they cannot handle. These top players in the Indian software industry are export-oriented firms with few inter-firm linkages within the clusters.

On the other hand, horizontal inter-firm linkages also do exist among domestic SMEs that work on product development. For example, a local small firm that I interviewed, which specializes in media-related hardware development, engages internally in upper-stream activities such as conceptualization, specification, and design as well as the final stage of testing/adaptation, but it subcontracts lower-stream activities such as coding and manufacturing/assembling to other local firms.

Indeed, many firms that I interviewed in Bangalore maintain that they have close social networks among themselves like "friends," characterized by reciprocal relationships of cooperation.[13]

Given the more or less exclusive focus on the exports market, and as each firm has exclusive contractual relationships with foreign firms within the cluster or abroad, domestic firms in Bangalore do not directly compete against each other, because they are mainly serving the US customers. As a result, these social networks promote cooperation, rather than competition, among local firms.

Since the 1950s, many state-owned research institutes, universities, and R&D facilities related to the defense and aerospace industries have been concentrated in Bangalore. This concentration in turn helped Bangalore accumulate industrial engineering skills decades before it emerged as a leading software cluster (Heeks, 1996; Parthasarathy, 2000; Okada, 2005).

5.4.2 Technological and Organizational Capabilities of Domestic SMEs

In this subsection, I discuss various dimensions of firm-level technological and organizational capabilities among small- and medium-sized software firms in the Bangalore cluster. I focus on four dimensions: i) individual technical skills; ii) business models and strategies; iii) technology and innovation capabilities and iv) marketing skills.[14]

Individual technical skills   Preliminary observations of about 20 small- and medium-sized software firms that I interviewed in Bangalore suggest that the level

---

12   Infosys has set up a subsidiary firm within the same premises which looks after its BPO operation.

13   Interview at a small software firm in Bangalore (November 29, 2004).

14   These dimensions of firm-level capabilities build on six dimensions that Veloso et al. (2003) identified in their analysis of firm-level capabilities in the software industries in India, China and Brazil: a) individual technical skills; b) revenue model; c) technology; d) marketing skills; e) process maturity and f) management. Discussions in Veloso et al. (2003), however, are general and largely refer to leading software firms in these countries.

of individual technical skills of their employees, particularly in programming, is generally good, but lower than that of their counterparts in large Indian firms and MNCs. Due to the fierce competition among software firms in Bangalore to hire the "best talents," from leading engineering colleges such as the Indian Institute of Technology (IIT), smaller firms end up attracting the second-tier talents (for a detailed discussion on hiring practices among large software firms, see Okada, 2005).

The skill profiles of employees differ depending on their business models and types of activities. Those SMEs specializing in software application services and ITES rarely employ engineers with doctorates. Instead, many of their technical employees hold bachelor of engineering degrees in computer science and related disciplines. Quite a few of the small firms specializing in web development services commented that for such work, academic backgrounds do not matter much, suggesting that relatively low skill levels are required in software services.

On the other hand, owners of some new start-ups hold doctorates in engineering from well-recognized engineering universities such as IIT; they also have a strong sense of entrepreneurship and a keen interest in technological development. For example, a small firm with 30 employees, which specializes in embedded software for multimedia products (similar to iPod), was set up by a team of three engineers who hold doctorates from IIT. Their highly specialized technical skills have led them to start up this business, often specializing in higher value-added activities than low-end application services. Some had worked with multinational firms, before starting their business as a spin-off. Many of their employees hold master's degrees, such as the three-year Master's in Computer Application (MCA) or a Master's of Science (MSc.) in computer science. These firms easily find suitable employees, as over 100 engineering colleges are located in Bangalore and its surrounding areas (Okada, 2005).

Owners of new start-up firms are relatively young people, often under 30, who graduated from local engineering colleges and have worked for several years at large Indian software firms or global IT firms, particularly in the US. They often set up new firms with a few classmates from engineering college. The knowledge, skills and networks that they brought from their work experience abroad are important assets for these new start-ups. Bangalore's software cluster enjoys these close social networks among these former classmates from engineering colleges and their friends, who often exchange and share information and knowledge, and help each other when they encounter problems.

Software SMEs also invest in a good deal of in-house skills training for their employees, particularly in terms of technical and process skills. These training opportunities also help them retain their employees. As firms try to move up the value chain by moving from simple application work to more complex software solutions and consulting, they require their workers to develop domain knowledge and skills. Thus, they often offer training opportunities to their staff, often when they form a new project in a particular subsector (such as finance, banking, healthcare, and telecommunication). Some firms said they send their employees to training institutions to acquire and develop particular technical skills and foreign language skills (especially Japanese and German). On the other hand, firms that carry out BPO/ITES activities provide extensive training to their staff with more focus on soft skills and attitude training.

Business models and strategies   Various business models co-exist among SMEs in the Bangalore software cluster. Many SMEs rely for their primary revenues on services outsourced from abroad, particularly offshore software application projects for individual clients on a fixed-term basis. For on-site services, they tend to bill on the basis of staff time spent at clients' firms, but in offshore services, they usually receive fixed amounts, which they set out in advance in contracts. Thus these two kinds of services have provided them with relatively secure sources of revenues and ensured sustained growth, given the growing worldwide recognition of India as a major destination for outsourcing work. As many firms are still young, often only a few years old, their scope of activities is limited. Interestingly, however, many small firms I interviewed noted that they follow the successful business model of Indian software giants such as Infosys and Wipro. While these two giants have no contractual relationship with domestic software SMEs, they do play a mentoring role for local SMEs, directly or indirectly. This is particularly important for Infosys, which itself began as a small firm.

Firms often find it hard to differentiate their services and to grow beyond a certain level due to decreasing returns to scale (Veloso et al., 2003), but small software firms still enjoy growing opportunities in the global market. In particular, they often see BPO as a strategy to ensure quick revenues and easy success, because they can effortlessly staff their programs by scaling up easily replicable work (except call centers). While many firms have recently enjoyed rapid growth in BPO, the more technology-driven firms deliberately stay away from BPO services. This is partly because the BPO business model is very different from that of software development, partly because BPO services have a lower profit margin (US$12 per hour in BPO, compared to US$24 per hour in software development), and partly because BPO requires a high level of initial investment. Indeed, a BPO firm noted that its infrastructure and utility costs are generally three times higher than those of software firms, due to its 24-hour operation.[15]

Few small firms are engaged in developing software products for several reasons. First, the entry risk is high, involving high development costs and a considerable waiting period before their profits appear. Second, without an internationally recognized brand, it is hard for Indian firms to export software products and packages. Third, the sale of products requires their continuous physical presence in the market to provide after-sale support to their customers, which is very costly, indeed, beyond the reach of many small producers.[16] Finally, many software SMEs

---

15   Interview with a manager of a BPO firm in Bangalore (December 6, 2004).

16   Interview at a small domestic Bangalore firm specializing in web technologies (December 2, 2004, Bangalore).

are self-financed firms, making them risk-averse; few are financed by venture capital, which still plays a very limited role in the Indian software industry. Small firms may be interested in product development in the long run, but only a small proportion of them are actually engaged in such activities for revenues, mainly for the domestic market, particularly targeting SMEs as clients. Some small firms do engage in product development in a long-term strategy to diversify their scope of activities, while mainly performing the more profitable application services to guarantee a revenue stream.

However, many SME owners are aware that the current services-oriented business model will not last very long for Indian firms, as the nature of competition in services, particularly for relatively simple application work, is essentially price-driven. That is, when wages go up in India, these firms will lose their competitiveness in the global market as the work moves to companies in lower-wage economies. They are acutely aware that they need to move up the value chain, to higher value-added activities.

Thus, some owners of new start-up firms began to develop their own products after first developing software solutions services for multiple customers, gradually extending their services as packaged products. Some other SMEs that had focused on applications development have recently started opting to merge with foreign (mainly US) business consulting firms to offer more consolidated offshore services in business and software solutions.[17]

Technology and innovation capability   In India, the level of technology penetration is generally low, compared to more advanced markets. The exception is software development, which firms conduct for US customers. Among small- and medium-sized software firms, those that are more technologically oriented (often those interested in product development) tend to invest in applied research as part of their R&D activities, albeit small, in order to promote innovation and expand the features of their future products. These R&D activities include prototyping, requirement analysis and product conceptualization. In some cases, domestic SMEs are engaged in R&D activities subcontracted from MNCs located in Bangalore, such as Motorola and TI, in terms of design, prototyping and testing. Some firms are even involved in pure research, though on a small scale.[18] In addition to in-house R&D activities, software SMEs acquire new knowledge and technologies in many ways: through their foreign clients, through joint work with local universities (e.g., their former academic advisors) and research institutions,

---

17   For example, Eximsoft (with 130 employees) merged with Trianz (US consulting firm, with 140 employees) in 2005.

18   Interview at a small Bangalore firm with 33 employees, which specializes in software applications and product development (November 30, 2004, in Bangalore).

and through the Internet. They also learn as they work with local institutions such as STPI and ESC, large MNCs, and through mergers with foreign firms.

On the other hand, more service-oriented firms, particularly those focusing on BPO, invest less in technological development. They mainly use Microsoft or Oracle platforms, and thus their own technological upgrading efforts are less intense. However, even these service-oriented software firms must themselves keep up with the latest technologies so they can add more complex features in application work to meet customers' demand. Thus, they carry out applied research mainly for process innovation that will improve their efficiency in business processes; they engage in requirement analysis, solutions development, implementation and integration, prototyping and learning new technologies. Just like larger firms (Veloso, et al., 2003), they want to ensure that their process capability is acceptable to foreign customers, so they are keen to obtain internationally recognized certifications of process maturity such as the Software Engineering Institute (SEI)'s Capability Maturity Model (CMM)-Level 5, which is the highest level.

They mainly acquire new knowledge and technologies from their customers, from the open market, from employees who have worked abroad, and by paying licenses and technical fees. They also learn new ideas from published reports and professional papers. Much of their learning entails their understanding of both the latest technologies and what their competitors do in the market. Thus, they largely develop their innovation capability through a daily process of learning and experience. Their foreign customers' requirements are an important motivation to develop their capability.

Marketing skills   SMEs face difficulties in marketing their products, partly because they lack an internationally recognized brand. Service-oriented firms often market their services through their own marketing offices, agents and affiliates located in their main markets (notably, the US, Europe, and to a lesser extent Japan, Australia and Singapore), regardless of whether they work on projects or on long-term consulting. These SMEs tend to expand their business through these intermediaries, and through references and personal networks (friends and relatives abroad, who are often NRIs), suggesting that social capital plays an important role in marketing and expanding business opportunities. These firms are interested in developing their own brand in foreign markets, rather than continuing as subcontractors to overseas clients. In fact, in the 1990s, more than 250 of the leading Indian software firms established their subsidiaries and branch offices in the US (NASSCOM, 2001). Small firms are now taking the same step. These intermediaries also serve as an interface; they translate customers' requirements and quality standards, and transfer technological and domain knowledge as well as information on the market to Indian SMEs.

Like the Indian software giants, even the software SMEs are keenly interested in diversifying their export markets, which are currently highly concentrated in the US. Thus, these firms have increasingly promoted training for their staff in language skills (particularly in Japanese and German) and skills related to foreign business cultures and business practices.

5.5 Conclusion

In this chapter I have examined the role of SMEs in the development of India's software clusters, especially in Bangalore, and the strategies they use to develop capabilities to innovate and build their global competitiveness.

In the Indian software clusters, a handful of large domestic firms and MNCs have played a key role in driving the growth of both the industry and its exports, the number of software SMEs has increased in recent years. They are usually located in single or multiple operations in a single cluster. Some small firms specialize in high-end activities such as embedded software and IC chip design, including highly-specialized niche areas, but many SMEs are engaged in relatively simple software application development.

This study finds that software SMEs in Bangalore undertake a wide range of activities with different institutional forms. First, MNCs extensively use local small firms as designated subcontractors. They offer their designated subcontractors various opportunities, including seminars and training, and the subcontracting firms remain loyal users of these global players' products. Thus, although such subcontracting work does not require these SMEs to have high technological capability, it does offer them access to the new set of knowledge, skills, and information that these MNCs provided.

The study also finds that domestic software SMEs in Bangalore use various strategies to develop their firm-level technological and organizational capabilities. First, regarding individual technical skills, while domestic SMEs have no problem recruiting employees with skills necessary to do the work, their employees' level of individual technical skills is generally lower than that of employees in large domestic firms and MNCs. At service-oriented SMEs, the majority of their technical staff have bachelor of engineering degrees in computer science and other related fields and a few hold doctoral degrees. However, the more technologically oriented SMEs, which were often new start-ups established by a team of engineers, have employees with higher technical skill profiles.

Second, regarding business models and strategies, domestic SMEs follow various business models. Although few firms carry out product development as a main revenue source, some SMEs use a combined strategy: product development as a long-term investment and application development services as their short-term revenue stream. Some firms exclusively focus on BPO/ITES, whereas the more technologically-oriented SMEs tend to deliberately avoid entering into BPO/ITES and move into product development and a narrow range of niche specialties. Some firms have recently started opting to merge with foreign business consulting firms to offer more consolidated offshore consulting services such as business and IT solutions.

Third, with respect to technology and innovation capability, while the level of technology penetration is generally low in the Indian software industry, technologically-oriented SMEs tend to invest in applied research as part of their R&D, to promote innovation and expand the product features they can offer in the future. On the other hand, more service-oriented firms tend to invest less in technological development. To gain international recognition, some SMEs are even keen on obtaining internationally recognized certification of process maturity such as CMM (especially Level 5, the highest). In order for SMEs to build global competitiveness, it will be critical to enhance their technological and innovation capabilities.

Finally, regarding marketing skills, SMEs generally face difficulties in marketing their products, due partly to the lack of an internationally recognized brand. However, service-oriented SMEs often market their services through their own marketing offices, agents and affiliates located in their main foreign markets. These SMEs expand their business opportunities through these intermediaries, and through references and personal networks. As part of their efforts to diversify their export destinations, service-oriented SMEs are increasingly preparing their staff with foreign language skills and knowledge of business cultures.

However, Indian SMEs generally subcontract their outsourcing work through these agents, making it difficult to directly communicate with end users. Therefore, if SMEs want to move up the value chain, they will need to offer a complete package of consulting services, which allows higher value addition, rather than offering only a part of such services. SMEs will also need to improve the capabilities of those branches and agents, so they can negotiate directly with end users.

To sum up, Indian software SMEs have great potential to compete in the global market and play an important role in promoting both regional and national economic development. However, in order to expand their business opportunities, they will need to develop both their technological and organizational capabilities to build their global competitiveness. There is much room for public policy to support these small firms' efforts to build competitiveness, in terms of developing technical and marketing skills, and fostering technological and innovation capabilities.

References

Breschi, Stefano and Malerba, Franco (2001). The geography of innovation and economic clustering: Some introductory notes. *Industrial and Corporate Change*, 10 (4), 975–1005.

Breschi, Stefano and Malerba, Franco. (2005). Clusters, networks, and innovation: research results and new directions. In Stefano Breschi and Franco Malerba (eds), *Clusters, Networks, and Innovation*. New York: Oxford University Press, 1–26.

Bresnahan, Timothy and Gambardella, Alfonso (2004). Introduction. In Timothy Bresnahan and Alfonso Gambardella (eds), *Building High-Tech Clusters: Silicon Valley and Beyond.* Cambridge: Cambridge University Press, 1–6.

Centre for Monitoring Indian Economy (CMIE) (2004). *Indian Industry: A Monthly Review*, 2004 December. Mumbai: CMIE.

Cooke, Philip (2001). Regional innovation systems, clusters, and the knowledge economy. *Industrial and Corporate Change*, 10 (4), 945–74.

Electronics and Computer Software Export Promotion Council (ESC) (2005). *Statistical Yearbook 2003–04.* Delhi: ESC.

Government of India (2001). *IT Manpower, Challenge and Response: Interim Report of the Task Force on HRD in IT.* Delhi: Department of Secondary and Higher Education, Ministry of Human Resource Development, Government of India.

Heeks, Richard (1996). *India's Software Industry: State Policy, Liberalization, and Industrial Development.* Delhi: Sage Publication.

Humphrey, John, Mukherjee, Avinandan, Zilbovicius, Mauro, and Arbix, Glauco (1998). Globalization, FDI and the restructuring of supplier networks: the motor industry in Brazil and India. In Mitsuhiko Kagami, John Humphrey and Michael Piore (eds). *Learning, Liberalization, and Economic Adjustment.* Tokyo: Institute of Developing Economies, 117–89.

ILO—International Labour Office (2001). *World Employment Report: Life at Work in the Information Economy.* Geneva: International Labour Office.

Markusen, Ann (1996). Sticky places in slippery space: A typology of industrial districts. *Economic Geography*, 72 (3), 293–313.

NASSCOM (2001). *The IT Software and Services in India: 2001.* Delhi: NASSCOM.

NASSCOM (2002). NASSCOM-McKinsey Report. *Strategies to Achieve the Indian IT Industry's Aspiration.* Delhi: NASSCOM.

NASSCOM (2003). *Strategic Review 2003: The IT Industry in India.* New Delhi, NASSCOM.

NASSCOM (2004). *Strategic Review 2004: The IT Industry in India.* New Delhi, NASSCOM.

Nelson, R. R. and Winter, S. G. (1982). *An Evolutionary Theory of Economic Change.* Cambridge, USA: Harvard University Press.

OECD (2000). *A New Economy: The Changing Role of Innovation and Information Technology in Growth.* Paris: OECD.

Okada, Aya (2005). Bangalore's software cluster. In Akifumi Kuchiki and Masatsugu Tsuji (eds), *Industrial Clusters in Asia: Analyses of Their Competition and Cooperation,* 244–77. New York: Palgrave-Macmillan.

Okada, Aya (2004). Skills development and inter-firm learning linkages under globalization: Lessons from the Indian automobile industry. *World Development*, 32 (7), 1265–88.

Parthasarathy, Balaji (2000). Globalization and Agglomeration in Newly Industrializing Countries: The State and the Information Technology Industry in Bangalore, India. Ph.D. Dissertation, Berkeley, CA, USA: University of California, Berkeley.

Patibandla, Murali and Peterson, Bent (2002). Role of transnational corporations in the evolution of a high-tech industry: The case of India's software industry. *World Development*, 30 (9), 1561–77.

Piore, Michael J. and Sabel, Charles F. (1984). *The Second Industrial Divide*. New York: Basic Books.

Porter, Michael E. (1990). *The Competitive Advantage of Nations*. New York: The Free Press.

Pyke, Frank, Becattini, Giacomo and Sengenberger, Werner (eds) (1990). *Industrial Districts and Inter-firm Cooperation in Italy*. Geneva: International Institute for Labour Studies/ILO.

Saxenian, AnnaLee (1994). *Regional Advantage: Culture and Competition in Silicon Valley and Route 128*. Boston, MA: Harvard University Press.

Saxenian, AnnaLee (2001). Bangalore: The Silicon Valley of Asia? Center for Research on Economic Development and Policy Reform, Working Paper No. 91. Stanford, CA: Stanford University.

Saxenian, AnnaLee and Hsu, Jinn-Yuh (2001). The Silicon Valley-Hsinchu connection: Technical communities and industrial upgrading. *Industrial and Corporate Change*, 10 (4), 893–920.

Software Technology Park of India (STPI) (2004). *Performance of IT industry in Karnataka and Software Technology Parks of India-Bangalore*. Unpublished internal document. Bangalore: STPI.

Sridharan, E. (2004). Evolving Towards Innovation? The Recent Evolution and Future Trajectory of the Indian Software Industry. In Anthony P. D'Costa and E. Sridharan (eds), *India in the Global Software Industry: Innovation, Firm Strategies and Development*. Delhi: Macmillan.

Veloso, Francisco; Botelho, Antonio J. Junqueira; Tschang, Ted and Amsden, Alice H. (2003). Soliciting the knowledge-based economy in Brazil, China and India: A tale of 3 software industries. Unpublished report. Cambridge, MA: MIT.

Chapter 6

Understanding Incentives for Clustered Firms to Control Pollution: The Case of the Jeans Laundries in Toritama, Pernambuco, Brazil

Mansueto Almeida[1]

6.1 Introduction[2]

There is a strong concern about how the public sector can enforce that small and medium enterprises (SME) located in clusters comply with the environmental standards, tax and labor legislation without hurting the firms' competitiveness. *On the one hand*, it is a well-known fact that small firms' owners in developing countries face constraints (low level of education, lack of capital, focus on short-term profits, low-priced products, etc.) that make it difficult for these firms to comply with the labor, environmental and tax legislation (Dasgupta, 2000). *On the other hand*, there are examples of firms located in clusters that once they upgraded and started to comply with the environmental, tax and labor legislation, became more and not less competitive (Tendler, 2002). However, it is still not clear under what circumstances the public sector can push SME to comply with labor, tax or environmental legislation without harming firms' survival and competitiveness. In this chapter, I want to address this question by focusing on a program of pollution

1   Economist. Instituto de Pesquisa Econômica Aplicada – IPEA, Brasilia, Brazil. e-mail: mansueto.almeida@ipea.gov.br

2   The research and its interim findings grew out of an IPEA Project in 2004 and a MIT project started in mid-2006 and funded so far by the Brasília offices of the UK's Department for International Development (DFID) and the World Bank, in which Judith Tendler was Principal Investigator. Profound thanks go to the supporters of the research in the form of financing and continuing feedback on the proposal and the interim findings and papers: the DFID office in Brasilia, particularly Miranda Munroe and Ernesto Jeger, and the World Bank office in Brasília and to MIT's Department of Urban Studies & Planning, for its various forms of support for the project, including a stimulating environment in which to engage in discussions around this subject. The usual caveats apply here. The funders are not responsible for, or may not necessarily be in agreement with, the emphases that appear here. I also want to thank Judith Tendler, Salo Coslovsky, and Roberto Pires from MIT; Lenita Turchi from IPEA; Eduardo Noronha form UFScar and José Puppim Oliveira from Fundação Getúlio Vargas-Rio de Janeiro for their helpful comments in previous versions of this chapter.

control in a cluster located in the poorest region in Brazil, where for years informal firms flourished, local entrepreneurs did not cooperate and pollution was common. If we understand how this program succeeded under such adverse circumstances, we might draw some lessons to replicate this program elsewhere. This is the main challenge of this chapter.

There was a strong turn in the Brazilian government at the end of the 1990s toward promoting small and medium enterprises (SME) located in clusters[3]. In Brazil, policymakers, politicians and academics alike seem to believe that promoting small firms in clusters will necessarily trigger a developmental process with positive outcomes for poor regions and for local workers. But does any policy toward promoting small firms in clusters necessarily lead to economic, social and environmentally sound development? The answer is no. Sometimes, the way policymakers choose to promote SMEs might harm instead of helping these firms to develop in a sustainable way. This seems to be the case of garment firms in Sulanca[4] at the state of Pernambuco in Brazil, where for years the government avoided enforcing labor, environmental and tax legislation. Since no taxes were collected, the government did not feel obliged to improve the local infrastructure nor to adopt developmental policies targeted at improving local firms' competitiveness. This kind of unspoken agreement was labeled by Tendler (2002) as the devil's deal[5]; a deal whose major outcome was to endorse firms to stay at the informal level, where competition is based mostly on cutting wages and avoiding paying taxes.

The peculiar point about this deal is that even at the end of the 1990s, when the state government decided to have a more active role in promoting small and medium firms in Sulanca, policymakers seemed reluctant to enforce the labor, tax and environmental legislation , fearing to disrupt the economic dynamism of the garment sector. In other words, even when the state officials in Pernambuco had to recognize that there was some economic dynamism in Sulanca, they were inclined to see this economic growth as the unique result of competitive advantages of not paying taxes and not complying with the labor and environmental legislation.

---

3   I will use the term cluster in this chapter in a very broad way. I mean by cluster an agglomeration of firms where there is a significant number of small firms producing the same product and these firms are located close to each other. This broad definition is quite different from the one used in the literature based on the industrial districts in Italy, but it is the most appropriate one to capture the way this term is used by policymakers in Brazil.

4   Sulanca is a region in the Brazilian Northeastern State of Pernambuco, 150 km away from the capital of the state, where there is a cluster of garment firms formed by three cities: Toritama, Santa Cruz do Capibaribe and Caruarú.

5   According to Tendler, J. (2002), the devil's deal involves an unspoken agreement between politicians and small firms' owners. This deal works in the following way: "if you vote for me, I won't collect tax from you and I will not make you comply with tax, environmental and labor legislation. The interesting point about this tacit deal is that, in many cases, politicians and policymakers alike think that they are truly helping SME to survive.

This is also the common view among many economic development officials who work with SMEs in Brazil in lagging regions. They tend to believe that burden-relieving policies (tax exemptions, subsidized credit, etc.) are the right ones to promote SMEs that produce low-quality products in regions where cooperation is not common, and the local labor force lacks schooling and training. Therefore, the unspoken agreement of not enforcing tax, labor and environmental legislation in exchange for political support named by Tendler (2002) as the devil's deal might work under two quite different circumstances. In one case, it might reflect the absence of the state in supporting the economic development of a region and, in the other; it takes the form of a second-best strategy to promote SMEs in lagging regions.

In this last form, policymakers claim that it is better to have some local economic growth going on than having to face the local economic disruption that would result from trying to enforce tax, labor and environmental laws on firms that barely produce to survive. Since policymakers focus on the "burdens" (the costs of formalizing and observing tax, environmental and labor legislation) themselves as the source of the problem, they advocate reforms that grant special relief from these burdens to small firms in the form of exemptions from taxes, and from labor and environmental codes (see Tendler, 2002:. 3).

Once we recognized that in many places policymakers see the devil's deal as a second best strategy to promote clusters, we need to understand why the government should have a more active role in supporting SMEs that sell to low-quality demand markets. Some entrepreneurs in poor regions or countries intentionally target low-quality markets and try to "innovate" in product and process to reach these markets, in which they perceive a strong opportunity to succeed. Under these circumstances, policymakers find it extremely hard to push these firms to innovate to sell to more high-quality demanding markets and it is not even clear if the government should push firms to move from one strategy (selling to low-quality demanding markets) to another (high-quality ones). But the case of garment firms in Sulanca is not that simple. In Sulanca, there is a problem of coordination similar to the prisoner's dilemma in the game theory[6]. Since the majority of local firms do not comply with tax, labor and environmental legislation and the economic policy is designed to support this strategy, informality becomes the most attractive option. The challenge for the government is to push firms to go from this bad equilibrium to the good one, where competition is based mostly on innovation; firms comply with tax, labor and environmental legislation and get in exchange a higher provision of public goods by the government.

---

6   The prisoner's dilemma describes a situation where two prisoners are known to have committed a crime and each are held in separate cells to be interrogated. The "dilemma" faced by the prisoners here is that, whatever the other does, each is better off confessing than remaining silent. But the outcome obtained when both confess is worse for each than the outcome they would have obtained had both remained silent.

Once we understand the right mix of incentives that leads entrepreneurs to cooperate and compete based on innovation instead of cutting wages, we need to ask how the government can change the incentives to push entrepreneurs to go to this high-road path. What are the factors under the direct control of the government that might lead firms to invest on upgrading and innovation? What kind of incentives must the government adopt that makes formalization the most attractive option to SMEs? How can the government push firms to comply with the environmental legislation without harming small firms' survival?

In this chapter, I want to answer these questions based on the case of the laundries in Toritama, one of the municipalities at Sulanca, where there are 2,000 small and medium firms producing jeans and 60 establishments that do the laundering for local jeans producers. The majority of these firms in Toritama are informal and for years the laundries avoided paying taxes, fringe benefits, and did not take any action to control water pollution, polluting the only local river in the city: the Capibaribe River. The pollution of the local river made it useless for local consumption. This situation has changed recently due to a set of economic and political factors. On the economic side, the soaring prices of water led the owner of the largest local laundry to seek for a technology to use recycled water to wash jeans. On the political side, the public sector got directly involved to push other laundries to do the same once a low-cost technology to control for water pollution was made available. In less than two years from 2003 to 2005, all the laundries in Toritama made investments to control water pollution, all of them turned to the formal sector and entrepreneurs started a business association.

How can we explain why the laundries, located in a place where informal firms flourish, local entrepreneurs hardly cooperate and pollution has been common, decided to invest in a new technology of pollution control? I claim that the answer to this question lies in three aspects: (a) the clear link between the benefits of upgrading (water recycling) and profits; (b) the development of a customized and less advanced technology that fits the need of small and medium laundries; and (c) the direct involvement of the government in enforcing the environmental law and actively helping firms to upgrade. All these factors together are important to explain why informal laundries in Toritama turn to the formal market, started to comply with both the environmental law and, partially, with the labor legislation. More striking, the new laundering technology adopted by firms in Toritama is quickly spreading to the other jeans cluster in the region (Caruarú).

In addition to this introduction, this chapter is divided into four sections. In the next section, I will show the pollution problem in Toritama and explain why the standard policies of pollution control would not work there. In the following section, I will focus on the history of the laundries in Toritama, trying to explain how these firms started to comply with the environmental law recently, stressing the three factors that made it possible for the laundries to upgrade. The fourth section explains the institutional cooperation within the public sector to foster the upgrading of the laundries. In the last section, I summarize the major conclusions of this chapter.

6.2 The Informal Sector, Pollution and Policy Options in Toritama

Toritama is a city in Sulanca, a region in the state of Pernambuco located 150 km away from the capital Recife, and is the major garment cluster in this state. The name "Sulanca" is a combination of "Sul", or South, where most of the cloth came from, and "helanca", which refers to the type of material (knitted cloth like lycra) used the most in the 1970s. The name in the 1990s had come to symbolize the cheap and low-quality product that dominated production at Santa Cruz do Capibaribe, Toritama and Caruarú, the three cities where the majority of garment production takes place.

Sulanca's image as a place where someone can find cheap products continues basically the same, but the stigma of "low-quality products" that has characterized the garment firms in Sulanca for years started to change. Nowadays, it has become common for middle-income consumers to travel from Recife to shop at Sulanca and, especially, in Toritama, a city of 20 000 inhabitants where 2,000 firms accounted for 15% of the jeans produced in Brazil in 2005 according to Serviço Brasileiro de Apoio às Micro e Pequenas Empresas (SEBRAE).

The last stage of producing jeans is the laundering phase, where the jeans is washed using from 60 liters to 100 liters of water for each pants or jeans[7], detergents, fabric conditioners and chemical products to get different tones of blue and colors. The daily consumption of water by a laundry in Toritama goes from 50,000 to 400,000 liters depending on the size. The effluent pollution in Toritama caused by laundries amounted to a biochemical oxygen demand (BOD) of 600 milligrams per liter of water[8], while the Brazilian environmental legislation requires a BOD of only 50 milligrams per liter.

Local entrepreneurs and state officials estimate that one million (1,000,000) pieces of jeanswear are washed each month in Toritama, which represents a joint consumption of 80,000,000 liters of water (or 21.1 million gallons) per month for all the 60 laundries in Toritama. Since the majority of the water used for laundering is transported by water trucks, this is equivalent to nearly 7,000 water trucks[9] per month loaded with water to wash jeans in Toritama. All this amount of water used to be discharged directly into the only river in the city, the Capibaribe river, without any previous treatment to clean the water from the chemical products used during the laundering, which made the contaminated water dark blue. Some local

---

7   All the technical information reported in this section comes from an interview with Jefferson Silva, a chemical engineer from the Pernambuco Institute of Technology (ITEP).

8   Biochemical oxygen demand (BOD) is the amount of oxygen required by aerobic microorganisms to decompose the organic matter in a sample of water, such as that polluted by sewage. It is used as a measure of the degree of water pollution. Also called *biological oxygen demand.*

9   A typical water truck´s capacity goes from 2,000 to 4,000 gallons of water. Taking an average of 3,000 gallons per truck, 21 million gallons of water would fit into 7,000 water trucks.

entrepreneurs used to say proudly that they were helping the flow of the river by releasing individually over 50,000 gallons of blue water daily into the river.

In addition to the pollution of the river, the laundries caused three other environmental problems. *First*, the chemical products used by the laundries were packaged in plastic packages, which used to be discarded directly in the streets. *Second*, the fuel used by laundries in Toritama to feed the furnace is wood. The use of burned wood to produce energy had two unwelcome environmental impacts: (a) the growing deforestation of the scarcely green fields near the city, and (b) the strong odor produced by the smoke released from laundries' chimneys. *Third*, the laundries did not have access to sanitary sewers and they did not have individual septic tanks, meaning that all the liquid and solid waste used to be discharged directly on the streets and into the river. *Finally*, the working conditions in the laundries were extremely poor and dangerous, especially for those workers operating the furnaces[10].

The continuing pollution of the river in a region where shortage of water is the rule together with the strong odors emanated from the smoke released by the laundries' chimneys pushed the population to demand some action from the new public state attorney appointed to the city in 2001[11]. The new attorney at the city, however, felt powerless to solve this problem. On the one hand, the local population was demanding control on the pollution originated from the laundries. On the other hand, laundries' owners were reluctant to take any action to control the pollution because the technology to do that was expensive and also because they had been in the laundering business for over 10 years without having to control the undesirable pollution. The key dilemma faced by the local state attorney was how to enforce the law without disrupting the economic dynamism of the city. This is not a simple dilemma to solve and many cases reported in the literature shows many programs adopted worldwide to enforce environmental legislation that ended up producing thousand of unemployed workers (Dasgupta, 2000) or failed to produce lasting results (Blackman, 2000).

6.2.1 Why Sanction and Coercion Would Not Work in Toritama

There are many different approaches for governments to enforce firms to comply with the environmental standards (see Dasgupta, 2000). One approach is to impose

---

10   One of the most common problems has to do with the lack of equipment and training of those workers who operate the furnaces. This person is supposed to work wearing special clothes to protect from the heat and from the fire in case of explosion, but no one received special clothes or adequate safety training.

11   The current state attorney at Toritama, Sérgio Gadelha, told me that it is very common in the Northeastern Brazil countryside for the local population to seek the new attorney nominated to a city to update him about the major problems of that city. That is exactly what happened to him, when he was nominated to be the state public attorney in Toritama in 2001.

on the polluter the cost of pollution (the polluter-pays-principle) and push firms to do a cost-benefit analysis. The industry decides to produce only if the benefits of producing and controlling for pollution outweigh the cost. This approach uses economic or *market-based instruments* to control the pollution[12].

A second way to enforce the environmental law is the *sanction-based approach* (or command-and-control approach). The government tries to enforce the environmental regulation by compulsion and coercion, setting fines for those firms that deviate from the environmental standards and issuing judicial orders to close down polluting industries. This approach was used extensively together with market-based instruments by developed countries in the 1970s. This approach is also known as the technocratic approach, since it assumes that the lack of compliance with the environmental legislation is solely a technical problem and firms have the ability to do the required investments for pollution control if proper coercion is set by the government.

The third approach to enforce the environmental law is named the *compliance-based approach*. In this approach, the environmental agencies recognize that some firms are not complying with the environmental legislation because the firms' owners do not understand the legislation; they perceive environmental investment as being unproductive, and the high-cost small and medium firms face to adopt existing technologies of pollution control. In this approach, the public sector adopts a proactive action of disseminating information about pollution controls and working closely with the private sector to reduce pollution with the least social and economic costs.

In general, in the last two decades, developed countries have increasingly evolved from using solely the command-and-control approach to combine the three approaches above. But many developing countries still rely on the sanction-based approach and try to enforce the environmental legislation only by compulsion and coercion. In general, the programs to control the pollution in these countries based only on sanction and coercion have failed.

In 1996, for instance, the Delhi government and its Delhi Pollution Control Committee adopted a wide program to control the pollution caused by firms of different sizes and in different sectors in Delhi (see Dasgupta, 2000). The firms were hit by a series of court orders requiring them to take three different measures: (a) closure and relocation outside the national capital of those industries classified as highly polluting; (b) relocation of nearly 90,000 firms from the places classified as nonconforming areas; and (c) joint construction of Central Effluents Treatment

---

12   It is interesting to note that in a wide survey about the use of market-based instruments (MBI) in pollution control programs in Latin America (LA), the authors conclude that, on paper, LA countries had everything in place to use MBIs. But institutional weaknesses (underfunding, inexperience, lack of political will, etc) limited the effective implementation of MBIs in pollution control programs. In other words, the problem in LA countries is not the lack of legislation but the lack of institutional capacity to enforce the law. See Huber, R. M., J. Ruitenbeek, et al. (1998).

Plants (CETP) by those firms that produce waste water. The results of this program were unpleasant: firms that relocated continued to use the same polluting technologies in the new sites where they had chosen to locate; 1,328 factories were closed and 125,000 workers lost their jobs; and only few CETPs were built. According to Dasgupta (2000), this program took the view that pollution in Delhi could be solved solely by sanction and coercion —the sanction-based approach. It failed to understand that small firms have distinguishing characteristics (short-term perspective, diversity of size and products, the use of cheap and not necessarily the most efficient technology, low level of education among owners and workers, etc.)[13] and constraints that need to be addressed in any program of pollution control. The plan in Delhi ended up creating a social problem and failed to push firms to adopt cleaner technologies.

In sum, there are many case studies that show that governments will not succeed to control the pollution caused by small-scale industries if they try to do that relying solely on enforcement and sanction; the command-and-control approach[14]. Policymakers need to understand the constraints under which small firms operate and try to address each of these constraints that might limit the adoption of green technologies. Next, I show that the plan to control the pollution in Toritama is succeeding precisely because the public sector is taking into account the constraints small firms face to invest in "green technologies". When confronted with the problem of pollution, the public officials in Pernambuco felt co-responsible for the problem and decided to work closely with local laundries' owners to find a common solution. In doing that, public officials designed a program that involves both penalties and incentives; and adopted a participatory approach to control the pollution.

6.3 The Upgrading of Laundries in Toritama to Control Pollution

The success of the public sector to push laundries in Toritama to comply with the environmental law was possible through a process that blends together three elements: (a) a clear link between upgrading and profits, once a private firm got involved in searching for a technology to recycle water in laundering jeans; (b) the public sector role in enforcing the law and helping firms to implement the changes (stick-and-carrot policy); and (c) the development of a customized technology that fits the need and constraints of the local firms, especially, small and medium ones.

By December 2004, all the laundries in Toritama had signed and presented to the state government a proposal to implement the process to control water

---

13   About the special characteristic of small firms in informal sectors see also Frijns, J. and B. V. Vliet (1999).

14   See Dasgupta, N. (1997), Scott, A. (1998); Frijns, J. and B. V. Vliet (1999); Blackman, A. (2000); and Dasgupta, N. (2000).

pollution. The largest ten laundries were ahead of the schedule to implement the process of pollution control. Every laundry had turned from the informal to the formal market; each had started partially to comply with the labor legislation[15], and the entrepreneurs had started a business association to demand further action from the local public sector in the provision of public goods in Toritama. In addition, in exchange for entrepreneurs' efforts to comply with the environmental legislation, the public state attorney mediated a proposal to improve the water system and to build a sewer system in the city.

6.3.1 A Clear Link between Upgrading (Pollution Control) and Profits

Toritama is a place where firms prospered targeting low-quality demanding markets. Informality was never a problem until recently, when successful firms decided to target the high-quality market segment and the economic growth attracted the undesirable attention of tax authorities. However, the major stimulus for local laundries in Toritama to upgrade came at first not from the punitive action of the government, but from the active search of a local entrepreneur for a low-cost technology to recycle water.

Mr. Edilson Tavares, the owner of the largest local laundry (Mamute), did not have any special interest in controlling water pollution. He was interested in an economic solution to the high costs of water, the most important input in a laundry. The water was not a problem in the past, since both the number of laundries and the population were smaller; meaning that demand for water was not too high even during droughts. However, the growth of Toritama (population and the number of laundries) made the lack of water acute during the drought in 1999 and the laundries' owners had to pay soaring prices to buy water transported by trucks from nearby cities[16].

Mr. Tavares was near bankruptcy and got involved with the Garment Association in Recife (SINDIVEST), which was making a strong effort to recruit new members in the "Pólo do Agreste" to the association. In one of the meetings of the association in Recife, Mr. Tavares asked for help to develop a process to recycle water in the laundering. At first, no one could help him, but the association put him in contact with the German institution BFZ[17], which was interested in

---

15 I mean by partially comply with the labor legislation the fact that at this stage, the laundries were required to solve quickly only the problem regarding the safety condition and the training of those workers who work directly managing the furnaces. The full compliance with the labor legislation is still due.

16 Entrepreneurs estimate that the price of water increases by over 50% in droughts.

17 BFZ – Training and Development Centers of the Bavarian Employers' Associations – is a business association with 1,800 employees, which develops many projects through partnerships in Central and Eastern Europe, China, India, Brazil, Mozambique and Mexico.

developing an economic program in Northeast Brazil and had donated some funds to SINDIVEST.

The BFZ representatives brought a group of German researches to visit the laundry Mamute in Toritama with the challenge to develop a technology to control for water pollution and to recycle water using only local resources. The researches succeeded in developing technology, which was less advanced than the existing one, but 70% less expensive because it used only cheap materials easily available in Northeast Brazil countryside.

The BFZ officials agreed to make the technology available free of charge to Mamute if its owner invited all the other laundries' owners to show how the process of water treatment works once it was implemented. Mr. Tavares agreed to do that, but he did not convince his competitors at first of the benefits in adopting the new technology. In fact, Mr. Tavares was accused by his peers as the one responsible for attracting the undesirable interest of the public sector and the state public attorney to the problem of informality and unsafe working condition of the laundries, since the public attorney became interested in knowing this new technology of water recycling and water pollution control.

It is important to notice that Mr. Tavares through his firm, Mamute, was the pioneer in implementing the technology developed with the support of the German institution BFZ not because of his environmental concern, but rather because he saw a clear link between the adoption of the new technology and higher profits. The water pollution control was a positive side effect in the search for a technology to decrease costs by recycling water.

The active search by private firms for technology to drive down costs is, however, a legitimate source of innovation and upgrading. Rodrick (2004), for instance, claims that in third world countries innovation is undercut not because the lack of trained scientists and engineers, but instead by lack of demand from its potential users in the real economy—the entrepreneurs. And the demand for innovation is low because entrepreneurs perceive new activities or new technologies to be of low profitability. The point of Rodrick (2004) is that if let by themselves, entrepreneurs will under-invest in the search for new technologies and new activities since they bear the full cost of their failure and have to share with others the benefits of one's discovery. It is based on this view that Rodrick (2004) states that "the industrial policy of the 21st century" should involve a public-private collaboration in this random self-discovery of new technologies and new profitable activities.

The active search by a private firm for a new technology to recycle water together with the support from BFZ to develop this technology fits into this process of self-discovery that blends together the private and public sector. I turn next to the role of the public sector in pushing the spillover of the new technology and the enforcement of the law.

6.3.2 The Direct Involvement of the Public Sector in Enforcing The Environmental Law and Helping Firms to Upgrade

A new public attorney was nominated to Toritama in 2001. Mr. Sérgio Gadelha de Souto was a young male in his 30s who was strong influenced by the education he received from the state attorney's office ("Ministério Público Estadual") about the importance of the environmental law for sustainable development. The state attorney's office in Recife, the capital of Pernambuco, has a special department responsible for educating attorneys about environmental issues, although the attorney has to work with many different subjects once he is designated to work in a city in the countryside. When Mr. Gadelha arrived in Toritama, he was shocked by the lack of compliance of local firms with the labor legislation and the environmental law.

At the same time, the state agency for environment and water resources (CPRH) had received a complaint about the pollution in Toritama and had sent a team to the city to investigate its extent. The CPRH staff did not know the number of laundries, neither the amount of water used by these firms in Toritama. The staff of CPRH had heard about the cluster of small and medium garment firms in that region, but never had they thought that a cluster of small and medium firms could be the source of a serious environmental problem. Therefore, at the same time that the new state attorney arrived in the city and got involved with the pollution issue caused by the laundries, CPRH had started an investigation to assess the extent of the situation in Toritama.

At the beginning, the new state attorney felt powerless to enforce the environmental law since he thought that small and medium firms could not comply with the legislation under the current technology. In fact, the public attorney together with CPRH considered relocating all the laundries away from residential areas to a specific industrial site to be settled by the city council, where a common effluent treatment plant would be built. However, the case of one local laundry (Mamute) that had implemented a process to recycle water changed the attorney's perception about the cost of adopting cleaner technologies instead of relocating.

Once the attorney came to know the low-cost technology implemented by Mamute and that the technology could be easily customized to small and medium laundries, he started together with CPRH and other institutions in Toritama (especially SEBRAE and SINDIVEST) an educational program to convince local laundries to control water pollution. Once their efforts proved to be useless, they decided to work together to enforce the environmental law as a means to start a dialogue with the laundries' owners.

In August 2003, with the back up of the local state attorney, CPRH promoted a wide operation to check whether laundries were complying with the environmental legislation. In just one day, 10 out of 60 laundries were closed and the laundries' owners opened a dialogue with CPRH and the public attorney to build a plan to comply gradually with the environmental law. The objective of both the attorney and CPRH with the inspections and the temporary closure of some laundries was not to order the laundries to install pollution abatement equipment in a short period or face closure. Their objective was to start a dialogue with entrepreneurs in order to debate the gravity of the pollution in the city and to find out how the public sector could help the firms' owners to make investments to control the pollution. In other words, the sanction strategy was used as a means to start a dialogue with the private sector to build together a common plan of pollution control.

The public attorney and the staff of CPRH signed together with each laundry's owner a plan named "Termo de Ajustamento de Conduta" (TAC) in which the firm's owner recognized that his firm was not complying with the environmental law and he would start to adopt the necessary steps to control for the emission of gases, to manage the solid waste, to build the septic tanks and to clean the water before discharged it into the river. When firms' owners signed this statement of confession and purpose, they received temporary permission to operate from 8 (medium and large laundries) to 12 months (small laundries) to implement the water pollution control, and they agreed to send a report in every three months to CPRH, showing the progress in implementing the investments required to control the pollution. If a firm fails to meet the schedule it had presented to the state, its temporary permission to operate will be canceled and the firm closed.

The state government and the public attorney also acted together to market the new plan signed with laundries' owners and to attract private firms to Toritama to sell the necessary equipment for the laundries to do the investments. The state attorney, CPRH, the Brazilian Office to Support Small and Micro Enterprises (SEBRAE) and other local institutions also organized many workshops with banks to show to entrepreneurs in Toritama the requirements to apply for investment capital. In addition, the local state attorney negotiated directly with laundries' suppliers of chemical products for them to collect back the plastic packaging of the chemicals used by the laundries. In doing that, the public sector helped the local laundries to control for the solid waste pollution with no extra cost to these firms.

However, the most important and the most difficult part of complying with the environmental law, the treatment of the water, was solved quickly by private firms and public agencies that acted quickly to replicate the technology developed by BFZ to sell to laundries in Toritama. Once BFZ-associated researchers developed and implemented the project with Mamute, SINDIVEST invited workers from the Pernambucan Institute of Technology (ITEP)[18] to go to Toritama to study the process installed at Mamute. ITEP's workers moved fast to learn the technology developed by the German researchers and they built a program with SEBRAE to promote the diffusion of the new technology among small and medium laundries in Toritama, once the state made clear that all the laundries would have to comply with the environmental legislation.

The cooperation between ITEP and SEBRAE led these institutions to implement SEBRAETEC in Toritama, which is a program of SEBRAE to subsidize

---

18   ITEP is a quasi-public agency linked to the state government. Its revenue comes from selling many different services to the private sector. See http://www.itep.br.

innovation by small firms. However, instead of subsidizing individual firms, any firm interested in getting subsidies to have a customized project to install the tanks to control water pollution and to recycle water had to join at least three other firms, despite the fact that the projects for each firm would be individual and tailored to the needs of each one.

In September 2004, one year after the operation carried out by CPRH to inspect firms' compliance with the environmental law, 50 laundries out of 60 in Toritama had signed and presented to the government the action plan to comply with the environmental standards[19]. By December 2004, all the firms had installed gas filters in laundries' chimneys to control for the emission of gases; ten laundries were ahead of schedule in controlling water pollution, chemical suppliers had started to collect and recycle the plastic packages and every laundry had adopted safety procedures for those workers responsible for operating the laundries' furnace and these workers had been adequately trained and certified to perform this task. In sum, the plan to comply with the environmental law was moving ahead as expected.

One important point about the role of the public sector in enforcing the law is the key role the local state attorney in Toritama had in building the cooperation between the public and private actors. Based on this fact, someone might ask whether the firms' compliance with the environmental law was the unique result of a brilliant young attorney's accomplishment. If this was the case, the upgrading of laundries in Toritama for pollution control would be less interesting than otherwise since it would result from personal characteristics of a young attorney who had a special interest in environmental issues. *But this was not the case.*

First, the attorney was aware of the importance to protect the environment because he received specific training on this subject in the "Ministério Público" and he was also demanded by both local population and his supervisor to pursue continuous action on this subject. *Second,* the local attorney in Toritama was acting together with CPRH and he was strong influenced by this state agency to push firms to control the pollution. CPRH had a key role at the beginning to convince the attorney about the problem of the pollution caused by the laundries and this public agency became responsible to check whether firms were meeting the schedule they had agreed to follow to install the equipments to control the pollution.

Last, one important point to understand is why CPRH and the state attorney's office decided to cooperate to push laundries to comply with the environmental law only recently and not before. In the case of CPRH, as I have showed above, this state institution did not know of the existence of the laundries and the pollution problem in Toritama. Only after CPRH received a formal complaint about the pollution in Toritama in 2000, did this institution started to evaluate the extent of the problem there. It seems strange that a state agency responsible for supervising

---

19   The deadline for all laundries to comply fully with the environmental law goes from July to September, 2005.

firms' compliance with environmental legislation did not know a famous cluster of garment firms well known in Northeast Brazil. But this lack of knowledge was common among policymakers in the state of Pernambuco, where for years the state government denied any direct support to firms in Sulanca, believing that those firms could not compete with low-priced Chinese products neither could they upgrade.

On the part of the state attorney's office, the former local state attorney did not take any action to enforce the environmental law because he was not aware of the extent of the pollution, since CPRH only conducted a wide survey in 2002 to measure the pollution in Toritama. Only after CPRH got involved with the pollution problem in Toritama did the state attorney's office in Recife start to look carefully at this problem, instructing the local attorney in Toritama to act to solve this problem. Therefore, the action to control the pollution in Toritama did not happen before because there was no formal evaluation on the extent of the pollution and no pressure over the local state attorney to focus on this problem rather than on the others in that city[20].

In sum, the diffusion of the technology to recycle and to clean the water used by laundries was made possible by the institutional cooperation between public (state attorney's office, CPRH, the state secretary of science, technology and environment—SECTMA, etc.) and nonpublic (SEBRAE, BFZ, SINDIVEST, etc.) agencies, and the cooperation among these institutions and private entrepreneurs. The state attorney's office acted together with CPRH to enforce the environmental law; these agencies had a double role in enforcing the law and in tutoring local laundries' owners on how to meet the environmental standards.

6.3.3 The Development of a Customized Technology

One point that we need to have in mind is that the technology to control for water pollution already existed in other places in Brazil. However, this technology did not fit the specific need of the laundries in Toritama for the following two reasons: (i) the laundries in Toritama are located far away from each other, that made it impossible to build a unique physical structure to control the joint pollution; and (ii) the technology available to control for water pollution adopted by the laundries in South and Southeast Brazil was too expensive for the small and medium laundries in Toritama[21].

---

20   Another fact that could have triggered the public officials' concern over the pollution problem in Toritama was the increasing media attention this cluster was getting at the beginning of the 2000s. For instance, a high ranked state official in Pernambuco told me that the governor of the state only came to know of the problem in Toritama when he saw a report about the garment cluster in this city on the news in one of the major Brazilian TV networks.

21   According to Jefferson Silva from ITEP, the most common methods of purifying water used by laundries in Brazil are the following three: activated sludge process, the use

Therefore, the problem was not the absence of a technology to control the water pollution but the absence of a cost-effective technology to control the water pollution caused by small and medium laundries located next to residential houses. This was the challenge presented by the owner of the laundry Mamute to the German researchers financed by the German institution BFZ.

The technology developed (adapted) at first by the German researchers and subsequently by ITEP is quite simple (see Figure 6.1). It consists of building three tanks to purify the water. The first tank, the *equalization tank*, receives all the water discharged from the laundering. The second tank, named *flocculation tank*, is made of either brick or fiberglass; this is the tank where the chemical process of water purification occurs. This tank is filled with aluminum sulfate and the water is agitated by a small motor to provoke the process of flocculation (the formation of large aggregated particles or flocs) and gravity sedimentation, in which flocs settle toward the bottom of the tank and the clean water stays in the upper portion of the tank. Once the wastewater is divided into clean water and sludge, the clean water pass though a filter made of brick, sand and stone, and it is ready to be discharged or recycled in the laundering process. The "dirty water" (the sludge) pass through a sludge filter and, then, it is discharged into the sewage system.

This simple technology developed by BFZ, ITEP and private firms to the laundries in Toritama was successful because it is cheap and can be completely customized to the size of each laundry. For instance, the number and the size of flocculation tanks to be built depend on the size of the laundry and the amount of water to be cleaned. If the laundry is too small, it needs to build only one small tank and install the filters. Second, if a laundry is medium and needs two flocculation tanks but lacks physical space, the tanks might be arranged either horizontally or vertically. Third, instead of buying pre-made fiberglass tanks, the tanks and filters can be made using cheap local materials (sand, bricks and stone) and the local labor force. In fact, the labor force is one of the most important inputs to install the process of water pollution control. However, since in Northeast Brazil the labor costs are the lowest in the country, the overall cost of the project turns out to be low.

The development of a new technology customized to small and medium firms in Toritama departs from the common sense of linking innovation with the development of a breakthrough technology. This is a case of developing a quite simple technology less advanced than the existing one, but a technology that fits the need of small and medium firms. In fact, the water cleaning process adopted by firms in Toritama is not as effective as the most advanced technology to clean

---

of hydrogen peroxide, and the production of ozone through an electrical discharge. The *activated sludge process* requires large tanks and a long time to clean the water (5 m³ of water requires 12 hours). The *hydrogen peroxide* is also used in water purification, but the cost (US$ 0.37 for m³ of water cleaned) of this chemical substance is highly prohibitive for small and medium laundries. At last, the process of *cleaning water with ozone* requires a setup investment of US$ 250,000 and a high consumption of energy in the process.

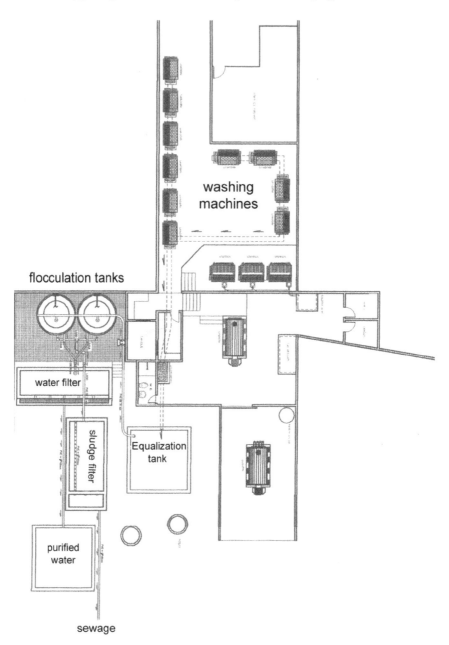

Figure 6.1   The plant of a laundry with the system to clean the water

water used by the laundries in South Brazil (activated sludge process, hydrogen peroxide, and producing ozone through electrical discharge). But this less advanced technology at low cost proved more successful than the others to reach small, medium and even large laundries in Toritama. In addition, the enforcement of the law by the public sector opened the market for private and public firms (ITEP) to step in and develop variant technologies similar to the one developed by BFZ to be sold to laundries at Toritama[22].

One might ask why local entrepreneurs have not searched for a low-cost technology to recycle water before, since the lack of adequate supplying of water is an old problem in Toritama. However, as I showed above, the shortage of water was a problem that became acute at the end of the 1990s, when the production of jeans increased at the same time that a long drought happened. The high prices of water led the owner of the largest local laundry to seek for a technology to recycle water, which had the positive side effect of controlling water pollution and becoming the benchmark for other firms and also for the public sector.

In sum, the process of upgrading was triggered by the self-interest of a private entrepreneur and the development of a low-cost technology. It is uncertain whether the public sector would have succeeded to enforce the environmental law, had not the private sector participated in this process of self-discovering to find a low-cost technology to recycle water. In addition, despite the economic benefits of cleaning the water to recycle, the role of public sector in enforcing the environmental law was vital to diffuse the technology adopted by the pioneer firm.

6.4 Institutional Cooperation and Upgrading

I have showed above that the state attorney's office together with the state environmental agency (CPRH) succeeded in pushing small, medium and large laundries in Toritama to make investments for pollution control. We identified three key characteristics that explain the success of this program: a) a clear link between upgrading and profits, once a private firm got involved in searching for a technology to recycle water in laundering jeans; (b) the public sector role in enforcing the law and helping firms to implement the changes (stick-and-carrot policy); and (c) the development of a customized technology that fits the need and constraints of the small, medium and large firms.

---

22   This case of developing a customized technology at low cost is similar to many cases of innovation reported by Prahalad (2005). In one of these cases, Prahalad describes the development of a prosthetic foot and lower limb prosthetics in India, the Jaipur Foot, using only local resources at low cost. The emphasis to innovate was placed on the cost of the materials used to build a low cost prosthetic that would fit the need of rural workers in India. Similar to the case of developing a low-cost process to control for water pollution in Toritama, in both cases, the development of this customized innovation at low cost was crucial to its success and diffusion.

In this section, I will develop further how the cooperation between public and nonpublic institutions evolved and the key role this institutional cooperation had in implementing the program of pollution control in Toritama. One reason behind the success of this public policy is linked to the fact that the institutions involved in this program went well beyond their formal duties to help local entrepreneurs to comply with the law. Instead of relying only on enforcing the law, the institutions in Toritama intensively helped local firms to implement the technology to control for effluent pollution, negotiated directly with chemical firms to collect back the empty chemical packages from the laundries, and actively supported firms' owners to organize and also to negotiate with high rank officials in the government[23].

In 2000, CPRH received a complaint about the pollution problem caused by the laundries in Toritama. The CPRH's staff decided to visit Toritama and check the extent of the problem. Before 2000, CPRH had never monitored the pollution in Toritama because they did not know about the existence of the laundries there and the huge volume of jeans produced in that city by small and medium firms. The CPRH's staff got shocked when they arrived in Toritama and saw the extent of the pollution, especially, the contaminated water the laundries were discharging direct at the Capibaribe River. The workers of this environmental agency never thought that a cluster of small and medium garment firms could cause an environmental problem that they expected to find only in the metropolitan region of Recife, where large firms from different industrial sectors are located.

The director of CPRH, Geraldo Miranda, then decided to meet different actors to try to elaborate a plan to push the laundries to control the pollution. At first, he met with the mayor in Toritama and some local politicians. Both were reluctant to take any action since they believed that laundries' owners could not comply with the law. Then Mr. Miranda met with the representatives of SINDIVEST, who were aware of the problem but felt powerless to take any direct action since SINDIVEST was just beginning a slow process of inviting garment producers in Toritama to join the association in the capital, Recife. At last, Mr. Miranda decided to meet with the state attorney in Recife and see whether the state attorney's office could help CPRH in designing a plan of pollution control in Toritama. As I have showed, the state attorney started a close cooperation with CPRH that resulted in the program explained in this chapter.

---

23   The proactive role of civil servants from the state agency responsible for pollution control was also essential to implement a successful program to control for industrial pollution in Cubatão, a city in São Paulo known in the 1980s as the "valley of death" because of the high levels of air, water and soil pollution. Lemos (1998) shows that the alliance between progressive technocrats from the state environmental protection agency (Cetesb) and popular movements in Cubatão was key to building a coalition to support the program to control the pollution in that city. See Lemos, M. C. M. (1998). Tendler and Freedheim (1994) also identifies the direct involvement of public healthy agents in performing tasks well beyond their formal duties as being a key factor behind the success of a preventive healthy program implemented in the Brazilian Northeastern state Ceará.

The director of CPRH decided to meet with the state attorney's office to begin a cooperation to solve the problem in Toritama because these institutions had already a history of working together that had started at the beginning of the 2000s when one of the directors of CPRH, Geraldo Miranda, was working in the city council in Recife and got involved in a program to control the pollution in garbage dumps. The pollution of garbage dumps in Recife is a well known environmental problem and also a social one since many poor families and children used to work collecting objects in these dumps. During the implementation of the program to set treatment plants in garbage dumps, Mr. Miranda came to know the branch of the state attorney's office responsible for enforcing the environmental law. Once Mr. Miranda moved from the city to the state government as a director at CPRH, he continued the program to control the pollution in the garbage dumps at the state level. In expanding this program, CPRH worked together with the state attorney's office to control the impetus of young state attorneys who were overreacting to the environmental problem in small counties, adopting a command-and-control approach and jeopardizing the program adopted by CPRH, which was more tutorial and based on incentives rather than on sanction and coercion.

Therefore, the cooperation between CPRH and the state attorney's office had started at the beginning of the 2000s and it was based on an effort to coordinate action to control the pollution in the garbage dumps. This early cooperation between these institutions opened a lasting communication channel between them that, eventually, evolved into the more recent cooperation to solve the pollution problem in Toritama.

The deep cooperation between CPRH and the state attorney's office is an important factor to explain the pollution control in Toritama. On the one hand, CPRH has the power to monitor a firm's compliance with environmental law, but lacks the power to order the closure of firms. On the other hand, the state attorney has the power to order the closure of firms, but he does not have resources to monitor firms' compliance with the law. When these two institutions work together, they can use a stick-and-carrot approach, helping firms to meet the environmental standards and use the threat of closure to trigger investments to control the pollution. But this partnership only worked because each agency truly helped each other and both went well beyond their formal duties. I turn next to this point.

6.4.1 Public Servant's Role: Going Beyond the Formal Duties

In the case of the pollution control explained in this chapter, the process to clean the water adopted by the laundries in Toritama is still insufficient to make the firms to comply with the Brazilian environmental law. The effluent pollution caused by laundries in Toritama amounted to a biochemical oxygen demand (BOD) of 600 milligrams per liter of water and this number is reduced by half once the process of cleaning the water is installed. However, the Brazilian environmental legislation requires a BOD of only 50 milligrams per liter. For the laundries to reach this required level of BOD, they will have to install an additional underground filter that requires a significant physical space not available to many small laundries. CPRH and the state attorney decide to accept the lower level of pollution instead of enforcing completely the law, fearing that a wide program that required the full installation of the system would fail[24]. In doing that, these two agencies had to reach a compromise; the state attorney had to accept that a pollution drop of 50% was the most that could be achieved in the short-run and CPRH had to agree to issuing a temporary environmental license to the laundries. Had either agency been inflexible in its duties, it is likely that the program would have failed at the beginning.

Second, CPRH not only cooperated with the state attorney's office but also persuaded the state attorney to pressure the state and city government to build a sewage system in Toritama. Once the meetings between CPRH, the state attorney and laundries' owners started in Toritama, the entrepreneurs complained about the requirement to install individual septic tanks in each laundry. Entrepreneurs claimed that the lack of a sewage system in Toritama was the fault of the government and, therefore, the state could not impose on them the burden of doing something that the government is supposed to do. CPRH officials agreed with the entrepreneur's view, but they felt powerless to pressure directly the high-level officials in the government. However, CPRH officials requested that the state attorney, who is in independent from the state government, do that. The state attorney monitored by CPRH had a meeting with the mayor and state's secretaries pushing them to build a sewage system in Toritama.

At the same time, CPRH officials informally fed the local press with news about the potential harm the pollution of the Capibaribe river could have on "Jucazinhos", the water reservoir that supply many cities in the region, including Caruaru, the most populated city in the region. CPRH officials also elaborated an internal report to the top rank officials in the government claiming that the risk of contamination of the most important water reservoir in Sulanca region (Jucazinhos) could have a negative impact on the political support of the governor's party in the region, since this reservoir supplies water to 83 cities in the region. Once this diagnosis was presented to the governor, he did not think twice and ordered that CPRH and state's secretaries get involved to solve the problem of pollution in Toritama.

The way CPRH officials framed the pollution problem in Toritama as being a potential political problem instead of relying solely on technical arguments was fundamental to push high-rank officials and the governor himself to support a solution to build a sewage and water system in Toritama. This kind of involvement of both CPRH officials, the head of the state secretary of technology and environment, and the local state attorney in taking sides with entrepreneurs in Toritama to push the state and the local government to build a sanitary sewage and a water system is far beyond the formal duties of these state workers. In fact, this

---

24   The requirement for the laundries to install this extra underground filter was postponed for a second phase, once each has completed installed the process to reduce water pollution.

broad involvement of public servants with their clients helped to build a sense of trust between CPRH, the local state attorney and laundries' owners that helped the implementation of the program of pollution control.

In December 2004, the state attorney, the mayor and the state government signed a plan —"Termo de Ajustamento de Conduta" (TAC)—in which the mayor promised to build a sewage system in one year and the state government promised to help the city's mayor to raise funds to finance the construction of the sewage system. In this same plan, it was established that the mayor will have to pay a fine if the sewage system is not built and CPRH will not renew the temporary permission for the laundries to run. In doing that, CPRH and the state attorney in Toritama pushed the entrepreneurs to pressure the mayor to build the sewage system, since the permission for them to stay on business will depend on the construction of this public infrastructure.

Lastly, the state attorney and CPRH officials together with other institutions (SEBRAE and SINDIVEST) strongly helped laundries' owners to build an association. Since the beginning of the public sector action in Toritama, state officials developed a strong partnership with the owner of the laundry Mamute, who helped the public and nonpublic institutions alike to schedule the meetings with the laundries' owners to convince them of the economic benefits of recycling water. The testimony of a local entrepreneur who enjoyed strong ties with laundries' owners and garment producers helped CPRH and the local state attorney to increase the dialogue and to earn the trust of local entrepreneurs to pursue the plan of pollution control. The increasing number of meetings to debate the plan to control the pollution also acted as a vigorous stimulus for entrepreneurs to start an association.

In the second semester of 2004, the entrepreneurs founded the Associação Comercial e Industrial de Toritama (ACIT). Once the public sector step in Toritama and entrepreneurs had to bargain with public officials, they had for the first time a clear incentive to mobilize and they actually received the incentive to do that directly from public servants. This is the kind of positive dynamic going from the public sector to the civil society, when civil society activism is in large a product of the initiative of the government or political parties[25]. In this case, public officials acted as "agents of change", working hard to foster participation and helping firms' owners to organize.

Although I am stressing in this section the cooperation between CPRH and the state attorney's office, it is important to point out that many other institutions were also important to trigger the process of institutional cooperation explained here. The Training and Development Centers of the Bavarian Employers' Associations (BFZ), for instance, had a key role in developing the technology used by Mamute and also in providing a course together with the Bavarian Ministry of Environment to entrepreneurs and state officials from Pernambuco in Bavaria. According to the director of BFZ, Martin Wahl, this course triggered a process of

---

25   See Abers, R. (1998) and Houtzager, P. P. (2003: 304).

institutional exchange among public and nonpublic institutions from Pernambuco that helped to increase the communication channels between them.

SEBRAE also had an important role in the process of educating entrepreneurs about the benefits and importance of controlling the pollution in Toritama. In addition, SEBRAE and ITEP implemented together the program to subsidize the design of the water treatment for small laundries. At last, but not least, SINDIVEST also had an important role in bringing BFZ and ITEP to Toritama and also in helping local entrepreneurs to associate. It is important to remember that the action of SINDIVEST to recruit new associates in Sulanca was what made it possible, at first, for BFZ in contact with Mamute, the laundry that was the first to install the equipments to clean and recycle water in Toritama.

In sum, the points raised above show that the institutional cooperation was key to help laundries in Toritama to invest on pollution control. In addition, the CPRH and state attorney's office went well beyond their formal duties to convince the government to improve the local infrastructure in Toritama and to help firms' owners to control the pollution and to build an association. This deep involvement of public servants with their clients is something that has been identified as a positive factor to explain the success of industrial policies (see Evans, 1995). However, this kind of embedding is usually absent in the modern literature of public administration[26], which suggests exactly the opposite; i.e., a clear definition of the tasks carried out by front-line workers to avoid these workers deviating from their core activities to do nonessential things. I will show next that the deep involvement of civil servants with laundries' owners in Toritama did not evolve to rent-seek activities because this close cooperation between private and public actors did not decrease civil servants' accountability.

6.4.2 *Outcomes vs Control: Public Sector Intervention and Accountability*

The modern literature of public administration assumes that once public servants have clear and limited tasks to pursue, it would be easier for high-rank officials to supervise "street level" workers. In addition, according to this view, the control over public servants to concentrate on their core tasks would decrease the opportunities for these workers to pursue self-interest. Based on this view, the efficiency of the public sector might increase when administrative reforms decrease public servants' discretion over their tasks and increase the control supervisors have over them. But many cases of successful developmental policies[27] happen when public and private actors are strongly linked to each other and public servants act in a very broad way. In fact, public servants' high discretion in South Korea, for instance, was important to support the industrial development despite historical cases of corruption of East Asian states before the 1960s.

---

26   See Tendler, J. and S. Freedhleim (1994) and the literature cited therein.
27   See Evans, P. (1995) and Amsden, A. H. (1989).

How can someone explain that the greater public sector involvement to promote the industrial development in East Asian countries did not ended up in rent-seeking activities? In other words, how did the state succeed in having a more active role to promote industrialization and, simultaneously, discipline business in East Asian? Amsden (1989) shows that the public sector succeeded in promoting and disciplining business in South Korea because subsidies allocation was tied to performance standards, where exports seemed to be the most efficient monitoring device. These kinds of policy, where incentives goes hand in hand with performance requirements, are stick-and-carrot policies. In the case reported in this chapter, the involvement of state officials with the private sector to enforce the environmental law help firms to upgrade and control the pollution worked because the higher discretion of public officials in pursuing their tasks did not decrease their accountability.

Northeast Brazil is a region where states have traditionally embarked in a fiscal war to attract industrial firms from South and Southeast Brazil (Tendler, 2000). The states give these incoming industrial firms fiscal and financial incentives, and usually finance labor force training. The problem with this policy lies in the fact that incentives are usually not linked to performance, and there is no evaluation of the net benefits these industrial firms bring to the local economy. In addition, the states usually do not make public the terms of agreement between the state government and private firms. Therefore, it is impossible for someone out of the government to check whether or not firms are complying with the targets they had set when they bargained with the public sector for fiscal and financial incentives. Neither is it possible to understand why some firms can renew incentives when they expire while others cannot.

The case of state intervention in Toritama to control the pollution is different from these traditional policies to attract industrial firms in Northeast Brazil because all incentives and requirements for each firm is clearly established in a legal instrument named "Termo de Ajustamento de Conduta" (TAC). This instrument works in the following way: once the state attorney's office finds out that a firm is not complying with the environmental law, the state attorney has two options. One is to firmly apply the law, ordering the closure of the firm and starting a lengthy and slow process in court that might take years to reach its final resolution[28]. A second option is to bargain with the firm's owner to sign a TAC, which is a document where the firm's owner recognizes that he is not complying with the law and agrees to follow a schedule to meet law standards. In this document, the state

---

28   This is exactly what happened in another cluster of jeans producer in Brazil Center-Western region. In the city of Jaraguá located in the state of Goiás, the public attorney tried to push laundries to comply with the environmental legislation based only on sanction and coercion. The laundries did not take any action to control the pollution, and the state and the laundries have engaged in a slow and lengthy battle in courts that might take years to reach a final solution.

attorney's office gives the firm's owners a grace period to comply with the law, and establish fines in case of no compliance.

The TAC is a highly discretionary device, since the state attorney has freedom to decide over the penalties and the extension of the grace period for each firm or group of firms. But despite being a discretionary device, each TAC is a public document published by the state official press. Therefore, any citizen can have access to these documents and check which incentives and penalties the state attorney's office and the state government established to make firm's owners to comply with the environmental legislation.

In addition, if any firm's owner fail to meet the targets established in the TAC, this document works as a personal statement of guilt, in which the entrepreneur assumes that he was violating the law. This confession speeds up the court process, making it easier to close noncompliance firms. In sum, the greater involvement of the public sector in enforcing the law and helping laundries' owners to upgrade to comply with the environmental law worked in a way similar to the stick-and-carrot policies adopted by East Asian countries. In the case reported here, the "Termo de Ajustamento de Conduta" (TAC) was essential to disclosure information over what the public sector had negotiated individually with each firm, the incentives (carrot) for complying with the law and the penalties (stick) in cases of noncompliance.

The use of the TACs in this program of pollution control was important to make the public officials accountable despite the high involvement public officials had with the firms they are supposed to inspect. Differently from the modern public administration literature, which overstates the control on public servants' tasks, the control in question was made on outcomes. In this case, higher discretion of public servants over their tasks did not decrease their accountability, and the use of the TACs increased the ability of the state to circumvent the historical delay of the court system in Brazil in case of prosecution. As the local state attorney in Toritama stated:

> we have tried to do this process in a very open and institutionalized way to make it independent from any individual. It does not matter with whom private entrepreneurs will be talking to one year from now, since everything that was negotiated with individual firms are set in each firm's TAC—a public document anyone can access to check whether firms' owners are meeting the conditions they had negotiated with the public sector.

6.5 Conclusions

This chapter analyzed a program still in progress to control for water pollution in a city in Northeast Brazil countryside (Toritama), where both the state and the local government used to be strongly absent. We have showed that the economic progress of this city comes from a cluster of jeans producers responsible for 15% of the jeans produced in Brazil. Despite the high informality rate, this city is also among the most dynamic ones in Northeast Brazil based on different economic indicators (high per capita income, low economic poverty rate, low dependence of governmental transfers, etc.).

The existence of a cluster of jeans in a region far away from the coastal area where the industry in Northeast Brazil still concentrates today is per se an interesting fact. The location of garment producers in Pernambuco's countryside and the resilience they have showed to compete with low-priced Chinese products is quite impressive. In addition, the location of jeans producers in a region famous for lacking an adequate water supply is something that challenges the comparative advantage theory and geographical barriers to development. Neither is this kind of economic growth explained by the new growth theories and its focus on innovation and education, which are usually characterized as being prerequisites for economic growth[29].

Despite the economic growth of the garment sector in Pernambuco countryside, I have showed in this chapter that this growth created an acute problem of pollution caused by the laundries in the jeans cluster in Toritama, a place where the government had been absent in enforcing the labor, tax and environmental legislation and also in providing public goods. The challenge of this chapter was to explain how laundries located in a place where informal firms flourish, local entrepreneurs hardly cooperate and pollution has been common, decided to invest in a new technology of pollution control.

I claim that the program to control pollution in Toritama is succeeding because of a process that blends together three elements: (a) a clear link between upgrading and profits, once a private firm got involved in searching for a technology to recycle water in laundering jeans; (b) the public sector role in enforcing the law and helping firms to implement the changes (stick-and-carrot policy); and (c) the development of a customized technology that fits the need and constraints of the local firms, especially, small and medium ones. Lastly, the role of the public sector in the case analyzed here involved three key elements: a strong cooperation within the public sector, a high dedication of public officials who self-enlarged their formal duties to help entrepreneurs to implement the program, and public-private cooperation to diffuse innovation developed within the private sector.

I also showed the double role the state had in Toritama—enforcing the environmental law and helping firms to upgrade—did not decrease the accountability of public officials. This happened because the private firms' owners' deal with the public sector was made public through a document named "Termo de Ajustamento de Conduta" (TAC), in which anyone can check the incentives and

---

29   I do not, however, dispute the view that education and innovation are important for development. What I am claiming here is that the focus solely on these factors as being the most important ones to foster development is misleading and might divert the public sector of adopting simple but effective growth enhance policies in poor regions. See chapter 4 of Easterly, W. (2001). for a critique about the focus only on education to explain economic growth.

penalties established by the public sector to allow firms to upgrade and comply with the law.

One point we must have in mind is that the case developed in this chapter happened in a place where the state had been absent for decades and local entrepreneurs did not cooperate. Therefore, the city of Toritama in Pernambuco's countryside is a place where we would not expect to find an example of a successful public policy, especially one that has to do with controlling pollution. The case reported here casts some doubt on the traditional way scholars have applied the well-known work of Putnam (1993). In Toritama, the lack of social capital was not a deterministic factor to deter the success of a program of pollution control.

The point developed in this chapter is closer to the idea developed by Locke (2001), who stresses that the government can be an important actor to help firms' owners to cooperate even under adverse circumstances (lack of social capital and lack of trust). According to Locke, cooperation within the private sector might arise when the public sector voluntarily push for cooperation in exchange for public support, there is self-interest of private actors to cooperate to get state's support, and there is the development of some mechanisms for self-governance and monitoring.

In the case explained here, however, the cooperation happened within the public sector and this institutional cooperation was a key factor behind the success of the public policy of pollution control. Self-interest of private actors in the case analyzed here also played a key role, but in a slight different way from that one explained by Locke. Here, self-interest is more linked to the private sector search for a low-cost technology to recycle water rather than on entrepreneur's incentives to start cooperating in exchange for public support. Eventually, entrepreneurs in Toritama built an association, but this happened later in the process.

At last, the jeans cluster in Toritama, a place where the public sector had been absent for years and a cluster where the state used to be ashamed, now turned into a successful case highly praised and publicized by state officials. Nowadays, instead of being a place the public sector tries to hide and exclude from state's developmental programs, Toritama is a place used by Pernambuco's policymakers as a showcase of their policy to support the development of clusters in the state countryside. In addition, as the secretary of science, technology and environment of Pernambuco, Claudio Marinho, stated:

> once the public sector stepped in Toritama to solve the problem of the pollution;
> it is just a question of time for the state government to start supplying health,
> educational services, and to have a more active role in supporting the development
> of the region.

References

Abers, R. (1998). From clientelism to cooperation local government, participatory policy, and civic organization in Porto Alegre. *Politics and Society,* 26(4), 511–37.

Amsden, A. H. (1989). *Asia's Next Giant: South Korea and Late Industrialization.* New York, Oxford University Press.

Blackman, A. (2000). Informal sector pollution control: What policy options do we have? *World Development,* 28(12), 2067–82.

Dasgupta, N. (1997). Greening small recycling firms: the case of lead-smelting units in Calcutta. *Environment and Urbanization,* 9(2), 289–305.

Dasgupta, N. (2000). Environmental enforcement and small industries in India: Reworking the problem in the poverty context. *World Development,* 28(5): 945–947.

Dohnert, S. (1998). Collective Services, Large Firms, and Clustering: Pulling Together the Threads of the Cearense and Pernambucan Garment Industries. Unpublished paper, Department of Urban Studies and Planning—Massachusetts Institute of Technology—MIT.

Easterly, W. (2001). *The Elusive Quest of Growth: The Misadventures of the Economists in the Tropics.* Cambridge, The MIT Press.

Evans, P. (1995). *Embedded Autonomy: States and Industrial Transformation.* Princeton, NJ, Princeton University Press.

Frijns, J. and B. V. Vliet (1999). Small-scale industry and cleaner production strategies. *World Development,* 27(6), 967–83.

Houtzager, P. P. (2003). From policentrism to the polity. In P. P. Houtzager and M. Moore (Eds.). *Changing Paths: International Development and the New Politics of Inclusion.* Ann Arbor, University of Michigan Press.

Huber, R. M., Ruitenbeek, J. et al. (1998). Market Based Instruments for Environmental Policymaking in Latin America and the Caribbean: Lessons from Eleven Countries. World Bank Discussion Paper no. *381.* Washington D.C.

Lemos, M. C. M. (1998). The politics of pollution control in Brazil: State actors and social movements cleaning up Cubatão. *World Development,* 26(1), 75–87.

Locke, R. M. (2001). Building Trust. Paper presented at the Annual Meetings of the American Political Science Association. San Francisco.

Prahalad, C. K. (2005). *The Fortune at The Bottom of The Pyramid: Eradicating Poverty Through Profits.* Upper Saddle River, USA: Wharton School Publishing.

Putnam, R. (1993). *Making Democracy Work: Civic Traditions in Modern Italy.* Princenton, USA: Princenton University Press.

Rodrick, D. (2004). *Industrial Policy for the Twenty-First Century.* Mimeo Harvard University.

Scott, A. (1998). The Enviromental Impact of Small Scale Industries in the Third World. Global Enviromental Change Programme Briefings (19).

Tendler, J. (2000). The Economic Wars Between the States. MIT/Bank of Northeast research project. Draft. Cambridge.

Tendler, J. (2002). Small firms, the informal sector, and the devil's deal. *IDS Working Paper* 33(3).

Tendler, J. and Freedhleim, S. (1994). Trust in a rent-seeking world: Health and government transformed in Northeast Brazil. *World Development*, 22(12), 1771–91.

Chapter 7

Social Technology for Mobilization of Local Productive Arrangements: A Proposal for Applicability

Ana Sílvia Rocha Ipiranga[1]
Maria Vilma Coelho Moreira Faria[2]
Mônica Alves Amorim[3]

During the last decades, several theoretical explanations have been set forth regarding the social and economic development models adopted in various regions of the world. Included in these are analyses related to industrial districts, *milieu innovateur,* clusters, local productive arrangements and systems. These approaches emphasize the role of agglomerations of companies specialized in the same products or services, located within a restricted geographical area.

Numerous studies have also highlighted the importance of localized agglomerations of small- and medium-sized enterprises (SMEs) for local development and mobilization of the local economies. In fact, several countries have recognized the SME's potential to create jobs and income and spark economic activity, contributing to a more equitable distribution of wealth, thereby compensating for regional differences. Various experiences in different countries, such as the industrial districts of the so-called "Third Italy", exemplify the successful regional agglomeration of SMEs, as do the clusters in the United States, led by Silicon Valley, Japan's, Korea's and Taiwan's entrepreneurial networks, and the local productive systems of France, Germany and the United Kingdom (Saxenian, 1994; Porter, 1998; Brusco, 1990).

In Brazil, small- and medium-size firms are of unquestionable importance, as 98% of the country's businesses are classified as small or medium size. These firms employ 60% of the economically active population and contribute to 21% of Brazil's Gross National Product (Ministério da Ciência e Tecnologia, 2001).

---

1   Professor at Ceara State University (UECE), Brazil. D.Sc. from the Università degli Studi di Bologna (Italy). Email: ana.silvia@pesquisador.cnpq.br.

2   Professor at the University of Fortaleza (UNIFOR), Brazil. Ph.D in Sociology from University of Tennessee, Knoxville, USA. Email: vmoreira@unifor.br.

3   Professor at the Federal University of Ceara (UFC), Brazil. Master in Planning (MCP), Massachusetts Institute of Technology. Email: monica_amorim@terra.com.br.

Despite their historical importance, it is recognized that Brazil's SMEs have not received the appropriate attention from the public sector, due to the erroneous perception that small businesses are inefficient organizations which, although contributing to the population's income, are not efficient enough nor sufficiently competitive to survive in a global economy.

SMEs are therefore considered almost as a panacea for diminishing the losses and solving social problems caused by unemployment. In this context, programs aimed at benefiting this segment are more of a social nature, as opposed to economically directed. SMEs are resorted to as a type of safety net against macroeconomic shocks, instead of being considered as efficient agents of production, capable of accelerating growth, improving income distribution, and obtaining competitive gains for the nation's economy (Tendler, 2002).

This has been the focus of recent policies in support of SMEs in Brazil, as well as in numerous other developing countries. As evidence of this, up to the mid-1990s, one of Brazil's main programs in support of SMEs was coordinated by the no longer existent Brazilian Assistance Legion-LBA, subordinated to the Department of Social Assistance.

The bases for including SMEs in social policies derives from the erroneous understanding that they are inefficient forms of production, and are therefore a "second class option" for a genuine strategy of economic growth. According to this viewpoint, nevertheless, these types of production units should not be left entirely to their own luck, as they have a favorable impact on political stability (*status quo*) by creating jobs and generating income for a significant part of the population, which otherwise would be idle and without access to economic activities, and by providing people with a start in business, while they are waiting for "better" options (Tendler, 2002).

This viewpoint is mistaken and may lead to significant losses of potential growth opportunities, especially considering that a development model based on SMEs makes it possible for more people to open small businesses, thereby profiting from their own production. The expansion of production associated with the enlargement of small entrepreneur segment constitutes a more just and socially desirable form of economic growth, without giving up efficiency.

Despite the unoptimistic viewpoint related to SMEs adopted by some countries, several others, such as Italy and Taiwan, have attained considerable growth figures over the last 20 years, mainly based on successful SMEs policies. In these countries, as well as others, SMEs make a relevant contribution to GNP and are treated accordingly by their governments, with support policies for the segment focused on production and linked to economic growth strategies, increase in the country's income, competitive gains and an increase in exports; all of which are essentially economic objectives, as opposed to mere social protection for the poor.

The experiences of the last decades clearly show that SMEs can play a crucial role in a country's development. Therefore, it is important to understand their origin and how they function, without comparing them to large companies. When SMEs are administered as independent production units, emulating large companies, they are unable to overcome their main difficulty, which is the inability to generate economies of scale. As these are typically associated with large volumes of production, which are almost exclusively attained by large companies, SMEs are relegated to a condition of inefficient production units, as their costs tend to be higher than those of the large companies, which reinforces the idea that SMEs cannot become important agents of economic development.

It is true that SMEs are affected by many problems, such as a scarcity of financing, unskilled labor, difficulty in acquiring new technology, etc. These difficulties are not easy to overcome, especially if faced individually by each of the SMEs (Tendler and Amorim, 1996). These businesses should, therefore, adapt their operations and organization to their production capabilities, insofar as the scale, scope and availability of production factors are concerned.

The need for efficiency and competitiveness forces the SMEs to develop their own organizational models, based on factors such as agglomerations, proximity, specialization and complementation. SME arrangement can encourage cooperation, learning, and the exchange of information among productive units, making them more efficient and competitive (Amorim, 1998).

Among the foregoing approaches, we have chosen to concentrate on "Local Productive Arrangements (LPAs)", as we believe they represent a more appropriate concept for the analysis of local productive agglomerations, especially SMEs, in less developed areas, and also because their analysis includes a network of agents interacting in a specific way, such as credit agencies, research and skill enhancement institutions, business associations and third sector organizations.

In this chapter, we consider local productive arrangements (LPAs) as "local agglomerations of economic, political and social agents, focused on a specific set of economic activities, working in conjunction" (Albagli and Brito, 2002:3). Complementarily, the so-called local productive systems (LPSs), representing more systematic methods, with a higher level of interdependence among the agents, are defined as "productive systems whose interdependence, articulation and consistent ties result in interaction, cooperation and learning, in turn resulting in the innovation of products, processes and organizational formats, and generating increased entrepreneurial competitiveness and social capabilities" (Albagli and Brito, 2002:3). Thus, the arrangement and local productive systems represent a more wide-ranging approach, as participants outside the productive activity are included, making them even more attractive, and offering the possibility of a deeper understanding of the dynamics of the economic, social and political surroundings of which they are a part.

This chapter focuses on local productive arrangements, as opposed to systems (LPS's), as the first are a better reflection of the situation of the productive agglomerations to be studied, which are, as of yet, in an initial stage of organization, without consistent coordination among the members and other local institutions.

Despite the rising level of interest in LPAs in Brazil, existing studies are more descriptive and/or analytical, as opposed to proposing and formulating strategies for productive arrangement development. In this context, a number of institutions (private, governmental, nongovernmental) have proposed Social Technologies (STs), aiming at formulating solutions to allow local actors in a productive territory to take advantage of local potential (Executive Department of Networks and Social Technology, 2004). This chapter intends to fill the gap of studies on local productive arrangements (LPAs) and their social organization, and hence proposes a Social Technology (ST) for Mobilization of Local Productive Arrangements, based on the concept of social capital (Putnam, 1993; Albagli and Maciel, 2003; Beretta and Curini, 2003), and using one of the three "ideal types" of governance proposed by Messner and Meyer-Stamer (2000), that is "network governance". Later, the application of this Social Technology for Mobilization will be exemplified by two cases of LPAs selected in the state of Ceará, Brazil.

We have used a methodology based on Bruyne et.al.'s (1991) approach, which presents the dynamic interaction among the four aspects of scientific research. These aspects are: the epistemological framework, in which the object of the research, and related problems are described; the theoretical framework, which guides the construction of the concepts, proposing rules for the interpretation of the facts; the morphological framework, which specifies rules for structuring the research object's format, by use of models, typology and "ideal-types"; and lastly, the technical framework, which controls the statistical research and the cross-checking of results with the theory upon which they are based. Based on the morphological framework, this study addresses the following questions: which instruments can be developed to stimulate the interaction, cooperation and coordination of the various actors involved in a LPA? In which way can these instruments be applied to the actual cases identified by the authors?

In order to answer these questions, we will first present a discussion covering the different typologies of SMEs agglomerations, focusing on social capital and governance, which are of strategic importance toward the sustainability and competitiveness of these types of organizational structures. Next, we consider the question of how to promote governance, including the three "ideal types" proposed by Messner and Meyer-Stamer (2000) and the model formulated by Humphrey and Schmitz (2000) on the three categories related to local and global levels. Based on the concepts of social capital and "network governance" and considering the methodological rules of the morphological framework proposed by Bruyne et al. (1991), in the third section we propose a Social Technology (ST) for the mobilization of Local Productive Arrangements (LPAs). In the fourth section, we present a possible application of the Social Technology, in two local productive arrangement (LPA´s), located in the state of Ceará, Brazil. These two cases are significantly grounded on local traditions and specialized on different products: handcrafted crochet and hammocks. In preparation for this study, we undertook exploratory research, which included participatory observation and interviews with "key sources" in different public and private institutions and businesses located in the two productive territories, as well as analysis of secondary data. In conclusion, this essay presents considerations on the possibility of applying the proposed model.

7.1 Typology of Firm Agglomerations

The new global economy's major characteristics are the development of regional blocks through the reduction of barriers among member countries, a greater use of information and knowledge, growth of the service sector, downsizing of large organizations and buyouts and alliances among companies. These factors, in addition to reducing job potential in the formal labor market, have stimulated the development of small businesses, especially those doing contract work for large companies (Lalkaka, 1997).

The change in organizational paradigms, brought about by the new global economy has resulted in the development of the flexible production model and new opportunities for SMEs efficiency, especially when they function in a collective way. As result of the productive restructuring of the 1970s, SMEs have begun to incorporate cutting-edge technology in their productive processes and to modify their organizational structures. They have also begun to seek links with the surrounding socioeconomic community, restructuring their operations in order to compete, in some sectors, with large companies. These transformations have occurred mainly in the technological innovations adopted by the electronic, robotic and computer industries. In this context, further debate is necessary regarding the importance of SMEs, given their successful results in creating jobs, even in periods of recession (Brusco, 1990; Castells, 1986; Piore and Sabel, 1984; Tendler and Amorim, 1996).

The economic globalization process has brought about a restructuring of the productive forces, emphasizing flexibility, innovation and quality. The expanding use of new flexible forms of production management points to the demise of the Ford model, with important consequences for the economic, social, organizational and technological spheres. The so-called "flexible specialization" stands out as a new organizational model, with successful examples from Third Italy, Germany, and France. These experiences were based on the expansion of small and medium size companies with adaptability to fluctuation in demand and capacity to innovate which made them important organizations in this new industrial model that has emerged in the last decades (Baptista, 2003).

The concept of "flexible specialization" leads us to yet another concept of industrial agglomerations, elaborated by Alfred Marshall (1996)— the "industrial district"—encompassing specialized firms and labor market, regionalization and cooperation. The resurgence of the idea of the "Marshallian District", during the recent period of production restructuring and exhaustion of the Taylor/Ford models, has caused an expansion of formal unemployment, increased informal work and aggravated conditions in the workplace. At the same time, the Marshallian Industrial District raises the possibility of forming a *locus* of cooperation, based on trust and on sociocultural aspects, and embracing networks of interaction among agents in a specific territory.

Despite the strong optimism that the industrial district model has provoked in face of the decline of classical methods of production management, the former has also shown signs of exhaustion. As pointed out by Schmitz and Nadvi (1999) and Le Borgne (1991), the Italian industrial district, a prime example of the Marshallian model, started to show changes in its operational dynamics in the 1990s. Cooperation has given way to subcontracting, emergence of leading firms, specialization in low-end products and increasing predominance of unskilled workers (Amaral, 1999).

There are several approaches to the study of firm agglomerations, such as: the new economic geography, with Krugman (1993) as its major exponent; cluster approach, led by Porter (1998); innovation economics, with Audrestch's (1998) leading contribution; and studies of small businesses and industrial districts, with important contributions from Brusco (1990), Schmitz (1994), Pyke, Becattini and Sengenberger (1992).

All these analytical approaches intrinsically apply concepts related to local productive systems, which have previously been used, in a more structured form, in developed countries, and more incipiently in developing countries, as it relates to regional and local development strategies.

The cluster strategy, as proposed by Porter, sits at the frontiers of two bodies of literature; the industrial organization and regional development, and demonstrates that sector analysis cannot grasp the complexity of the phenomena involved in industrial dynamics. Nevertheless, the cluster analysis attempts to grasp structural and systemic elements of firm agglomeration, emphasizing issues such as rivalry and internal competition. Thus, the analysis of business clusters prioritizes the study of their structure, investigating aspects such as: number of participants, level of interaction, standards of specialization and competition, and competitive advantages which can be developed with the structuring of this type of productive agglomeration. In addition to the foregoing approaches, studies of Regional Economics primarily emphasize aspects related to "location factors" which influence the establishment of firms in a specific geographical area, and the consequent effects on the reproduction and transformation of specific geo-economic regions (Britto and Albuquerque, 2003).

These approaches present some confluent and complementary points, as they emphasize the geographical proximity of productive agents and the relevance of the social and institutional context as important factors for the consolidation of these agglomerations. The cluster approach has more affinity with the large flexible production than with the small one. However, it differs from both the Ford model of production based on mass production in large companies, as well as the idea of the Marshallian industrial district with small, flexible production. In addition to emphasizing competition, rather than cooperation, this approach also suggests the importance of a key-firm or key-firms in a specific territory. While these anchor firms, undoubtedly, stimulate development of the territory through the mobilization of local productive agents; a more homogenous growth of productive agents becomes secondary.

In the face of diversity of theoretical framework for businesses' competitive performance, scholars seem to have reached a consensus that focus of analysis should not be centered exclusively on the individual firm, but mostly on the relationships among firms, and between them and the institutions which interact with them in a specific geographical area. This new focus has brought about changes in public policies for industrial and technological development. Therefore, the approach to arrangement and local productive systems proposes a model including traditional categories in the analysis of agglomerations, such as cooperation, but also including processes of interaction, learning, qualification and innovation; all these are increasingly considered fundamental for sustained competitiveness of agents participating in business agglomerations (Schmitz, 1994; Cassiolato and Szapiro, 2002). It then becomes relevant to consider the density of existent relationships among different agents of LPAs, in addition to the dynamics of social capital and governance as strategic categories for the sustainability and competitiveness of this type of organizational structure.

7.2 Social Capital and the Promotion of Governance

The interest in social capital is recent among scholars of organizational fields. Initially, discussions on the subject appeared among sociologists. Bourdieu (1985) disseminated the term in the 1980s. In this French scholar's view, social capital, as well as other forms of capital, is unequally distributed in society emphasizing the existent conflict of power in the social structure. Coleman (1988), an American sociologist, also stressed the interconnection between Sociology and Economics, pointing to the relationship between education and social inequality.

Later, Putnam (1993) turned the term "social capital" even more popular, with the publication of "The Community and Democracy: The Italian Experience". In this work, Putnam characterizes a community's social capital as a public resource that facilitates spontaneous cooperation, branching out in different forms and manifestations, such as, rules of reciprocity, a network of social relations, and a system of participation and trust.

Furthermore, Beretta and Curini (2003) define social capital as a "generalized expectation of cooperation", emphasizing, therefore, the "cooperation" dimension that affects the forms of coordination (governance) of the local productive agglomerations.

There are different types of cooperation in local productive systems and arrangements, including "productive cooperation", which aims at obtaining economies of scale, enhancement of quality and productivity levels, and "innovative cooperation", which allows for reduction of risks, costs and time and, mainly, promotes interactive learning, thereby favoring productive and innovative development (Cassiolato and Lastres, 2003).

In accordance with these perspectives, Albagli and Maciel (2003) emphasized that components of social capital favor innovation processes, interactive learning, in addition to fostering and sharing of knowledge, which are important dimensions to dynamism, while generating economic and social benefits, such as: sharing of information and knowledge; fostering a propitious entrepreneurial environment, contributing to the arrangement's competitiveness and survival; better coordination and coherence of action (governance) and collective decision-making processes, as well as greater organizational stability, hence contributing to cost reductions; greater mutual knowledge among the participants, reducing the risks of rent seeking, and favoring a greater commitment to the group.

According to Brusco and Solinas (1999), a complex *corpus* of rules — a code of trust — that regulates participants' behavior is the element that governs diffuse cooperation practices and enables participation in productive agglomerations of micro, small and medium size businesses. This code of trust, discussed at great length by Fukuyama (1996), involves not only issues related to life in the workplace, but also, in a more general way, the external conditions such as the associated life.

The development of this "code of trust", which regulates the participants' behavior in these networks of cooperation relates to the discussion of "governance promotion" in local productive systems and arrangements. As actions become less individualized and more group-oriented, it becomes necessary to create a coordinating force for the different activities in order to encourage collective actions and ensure attainment of desired objectives. Intensification of the relationship among participants and a better coordination of these relationships tend to facilitate promotion of governance, a necessary attribute for the LPA to evolve to a LPS. In this way, the emergence of "good" governance depends on a learning period which can be derived from the construction and consistency of the collective actions (Amorim, Moreira and Ipiranga, 2004).

Messner and Meyer-Stamer (2000) contributed to this discussion by emphasizing that governance in productive systems involves planning for negotiations among participating agents who are continuously interacting with themselves. Applying Max Weber's model, the authors classify three "ideal types" of governance:

1. The Hierarchical (modeled according to traditional instruments: money, power and the law);
2. The Marketplace (coordination based on the "invisible hand"); and
3. Networks (developed through negotiation, reciprocity and interdependence among participants).

In addition, Humphrey and Schmitz (2000) present three categories of governance (public, private and public-private), on a local and global level, each of them focusing on a specific relationship among agents operating in different areas of economic activity (Table 7.1).

Table 7.1   Categories of public-private and local-global economic activities governance

	Local Level	Global Level
Private Governance	Local business associations Hub-and-spoke clusters	Global buyer-driven chain Global producer-driven chain
Public Governance	Local and regional government agencies	WTO – Rules National and Supranational rules with global standing
Public-Private Governance	Local and Regional Policy Networks	International Standards International NGO campaigns

Source: Humphrey and Schmitz (2000).

These authors assume that the standard of governance based on networks brought about by negotiation is crucial for developing countries. The logic of interaction and styles of decision making in these networks tend to bring about learning processes, hence influencing the development of "social innovation". In fact, the discussion on governance has echoed in developing countries in which government and nongovernment institutions presented several proposals of Social Technologies (STs) aiming at developing a governance of productive territories. In Brazil, the debate on governance has also increased with the rise of several STs created to implement it on a local level.

The Executive Department of the Social Technology Network[4] (2004) argues that the possible meaning of STs seems an equally theoretical and practical challenge, as they go beyond a concept, encompassing ideas such as enterprises, associations of organizations, networks and attempts to foster cooperation. They also mean businesses, which generate jobs and income, and mostly, the recognition that merger of common knowledge and specialized knowledge provides powerful tools for social inclusion and human progress. STs then converge to the "social innovation" concept, given that the former constitute a process of innovation, to be carried out, collectively and in conjunction, by the participants interested in constructing a desirable scenario (Dagnino and Gomes, 2000).

Therefore, the concept of social innovation, taken from the innovation point of view—conceived as the group of activities involving from research and technological development to the introduction of new methods of labor management, with the goal of allowing a productive unit to create a new product or service for society—is a recurrent theme among scholars and ever more present in the sphere of public policies. This concept encompasses from the development of a machine (hardware) to a system for information processing (software) or

---

4   Secretaria Executiva da Rede de Tecnologia Social.

management technology—organization or government—of public and private institutions (orgware).

Dagnino and Gomes (2000) use the concept of "social innovation" in reference to knowledge—intangible or incorporated by people and equipment, tacit or codified—whose objective is to enhance effectiveness of processes, services and products related to satisfying social needs. That is, without excluding the previous interpretation, this one refers to a distinct code of values, development style, "national project" and objective of social, political, economic and environmental aspects. Similarly to the above, the concept of social innovation encompasses three types: hardware, software and orgware.

These discussions on the productive, institutional and social context which characterize the SMEs arrangement emphasize this important nuance of the innovation concept. Innovation is hence, not limited to introduction of a new productive process or a new product, but rather it also means development of new relationships and complex service capabilities, in which the product is only one component of the final supply (Rullani, 2003). It seems that a basic priority is the willingness to promote innovation in a comprehensive manner, creating an environment that facilitates the mobility of ideas used as conductors of innovation, promoting cooperation, either in the public or in the private and entrepreneurial sphere.

7.3 Social Mobilization Technology of Local Productive Arrangements (LPAs)

The Social Technology (ST) for mobilization of LPAs proposed in this essay is based on the authors' perception of a need to structure interventions aimed at promoting organizational arrangements such as LPAs in a distinguished way, compared to the traditional procedures used in the development of economic sectors, especially those used in the conventional approach to SMEs. The conventional approach emphasizes isolated actions, concentrating on individual firms, disregarding the group of participants of a specific productive arrangement and the joint actions that can be generated through the development of the existing social capital and governance, even if these are incipient. By proposing this technology, we intend to foster a discussion on collective and reticular forms of mobilization of local productive arrangements (LPAs), encouraging innovation and aiming at competitiveness enhancement and the evolution from an arrangement to a local productive system (LPS).

Considering the methodological rules of the morphological pole (Bruyne et al., 1991) and based on the concept of social capital (Putnam, 1993; Albagli and Maciel, 2003; Beretta and Curini, 2003) as social relation networks that facilitate cooperation and systems of participation, and also on the idea of "network governance" as proposed by Messner and Meyer-Stamer (2000), we propose a Social Technology for Mobilization, suggesting the development of three entities organized in a network. These entities constitute instruments of social mobilization and are based on an approach of cooperative learning intended to promote social innovation; as they depend on collective participation of participants interested in the strengthening of social capital and governance of the productive territory (Dagnino and Gomes, 2000). We begin by identifying a larger group of selected participants ("Forum for Change") which, due to the dynamics of its reticular structure, gets subdivided into various work groups ("Innovation Laboratories"), focusing on specific chores aiming at solving the LPA's problems. These work groups have mechanisms that explicitly link themselves to selected reference institutions ("Hearing Points"), which possess advanced know-how on topics especially relevant to the LPAs. The characteristics of these entities are described below.

7.3.1 Forum for Change

This entity focuses on facilitating spontaneous cooperation through which public and private institutional participants involved in the LPA will be invited to take part in a program for changes. The Forum incorporates the different forms of manifestation of the social capital described by Putnam (1993), such as the networks of social relations and the systems of participation. Considering that Cassiolato and Lastres (2003) emphasize the "innovative cooperation" which results from interactive learning, increasing the potential for the development of productive and innovative talent, the Forum functions as a meeting place for the exchange of ideas. It aims at ensuring a link between innovation proposals and those who will be called upon to manage and carry out the specific actions in each context.

In addition, the dissemination of these acts of cooperation (Brusco and Solinas, 1999), deriving from reciprocity, interdependence, planning and negotiation of a set of collective actions, tends to facilitate the establishment of a "code of trust" among the participants. Hence, it reduces rent-seeking risks, encourages a greater group commitment, and establishes a coordinating body which will help to promote a network governance among involved members (Messner and Meyer-Stamer, 2000; Albagli and Maciel, 2003).

The Forum should include the main participants interested in the development of the LPA. We take the example of the state of Ceará, located in Northeast Brazil to illustrate operational aspects of the Forum for change to support a LPA. The following institutions may be called upon to participate in this Forum (Table 7.2).

7.3.2 Innovation Laboratories

The strengthening of the social capital and the implementation of a coordinating body (governance), facilitated and articulated by the Forum, will tend to favor collective initiatives, such as interactive learning, knowledge sharing and innovation processes, development of an entrepreneurship prone environment,

Table 7.2   Composition of the Forum for Change

Composition of the "Forum for Change"
Selected Participants
1) Selected members of the LPA
2) Local government representative
3) Representative of SME Supporting Agency (SEBRAE)
4) Representative of Regional Development Bank (Bank of Northeastern Brazil)
5) State government representatives (Financing, Development, Employment Agencies)
6) Member of the Local Retailers Association – CDL
7) Representative of the Bank of Brazil
8) Representatives of Universities
9) Representatives of the Regional Institute of Technology (CENTEC)
10) Freight Carriers, Postal Service (SEDEX)
11) Representatives of LPA's Clients and Brokers

Source: The Authors

foster collective decision making processes and organizational stability. These are all of fundamental importance in promoting LPA dynamism as well as in generating economic and social benefits (Albagli and Maciel, 2003). In this context, we propose the establishment of Innovation Laboratories, targeted to key developmental issues and composed of operational work groups, whose members have shown distinguished performance with respect to the issue. Each Innovation Laboratory aims at developing solutions for a major challenge faced by the LPA, as well as implementing proposals discussed and decided in the Forum. The work group operates within a limited time period and should maintain systemic interchange with the other members of the Forum. The laboratories will have, therefore, different characteristics and modalities, but focusing on increasing effectiveness of processes and products, as well as on the development capacity for new interactions and for developing complex service projects (Rullani, 2003), including those related to satisfaction of social needs (Dagnino and Gomes, 2000). Based on specific needs, the Forum will institute new laboratories and discuss challenging matters, as they emerge (Table 7.3).

7.3.3 Hearing Points

Part of the literature on the subject stresses that development of LPA requires a dynamic balance of the degree of internal vitality of its fabric (internal synergy) and the external contacts (participation in global networks or external relations) (Ipiranga, 2006). Taking these discussions into consideration, some researchers

Table 7.3   Objectives of the Innovation Laboratories

Examples of "Innovation Laboratories" and their objectives
1. National and international observatory of new technological trends
2. Setting up a "pool" to prospect new markets, national and international distribution channels
3. Development of cooperative practices (relationships and exchanges) among businesses and among arrangements
4. Banking relationships and development of customized and collective forms of financing
5. Actions focused on the process of innovation, design, and diversification
6. Training and skills development initiatives
7. Development of new technical skills; Examples: design, logistics and marketing.
8. Management
9. Establishment of a network of partnerships with public authorities of the state, municipal and federal governments
10. Cohesion and interaction among various productive agents and institutions (national and international) for the exchange of best-practices
11 Social and cultural marketing initiatives aiming at enhancing the territory's image (intangible aspects);
12. Promotion of commercial actions among the SMEs and among the LPAs
13. Encouragement of Entrepreneurship
14. Relationships between contractors and subcontractors

Source: The Authors

(Maggione and Riggi, 2002; Blessi, 2002) have simulated different scenarios of interaction by means of the two alternative channels of collective learning— "the internal connections and the external connections"—in order to verify the implications on the process of creation and dissemination of innovation which emerge from the cultural, institutional, technological, productive and social fabrics in which the firms are located. In the case of the external connection, Maggioni and Riggi (2002) have found that external contact allows firms to learn and improve their performance. This contact happens only when a firm approaches a "technological window" for a period of time enough to observe, imitate and incorporate the innovation.

The objective of the "Hearing Points" is to function as a technological window to the outside world, functioning as sources of new ideas which can possibly be used by the LPAs. Each laboratory may have one or more Hearing Points, which should be carefully selected, depending on the nature of the productive activity and the problems it faces. It is important that members of the laboratories maintain permanent external contact and connections with each hearing point, in order to facilitate the interaction and access to new knowledge. It is not necessary for the members of the Hearing Points to be physically present in the laboratories, as contact can be made by virtual means. The Hearing Points may be traditional institutions, or simply references or virtual entities (e.g., web pages; businesses consortiums, trade fairs outside the arrangement).

In order to allow for a better understanding of the Social Mobilization Technology and its reticular organization dynamics, we present next a visual plan for its operational structure.

Figure 7.1 demonstrates that the three organizational entities are intensely interrelated, constituting a network of relations that strengthen productive and innovating interdependence. This allows for efficient collaboration with information flows, linkage of actions and gains in learning and productive efficiency. Based on this interrelationship and to the extent that discussions and negotiations advance among participants, definitions on the set of actions to be implemented through the Mobilization Technology become clear, allowing for cooperative learning and social innovation (Dagnino and Gomes, 2000). During the entire process, management of the Social Technology (ST) is achieved by means of instruments and empirical charts elaborated and implemented by participants involved in each stage, such as evaluation and impact assessment and implications of the use of this technology in the territory. In sum, to the extent that it facilitates interrelationships, learning and

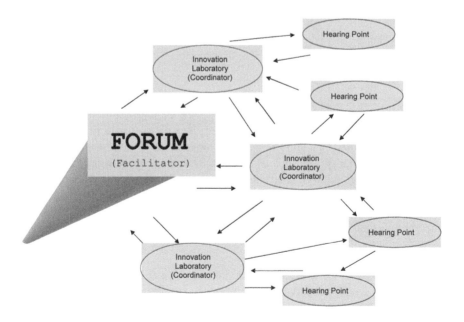

Figure 7.1   Social mobilization technology through a cooperative approach

innovation, development and practical experience, this interrelationships network may become an effective mechanism to help the LPA to evolve to a more complex structure, such as the SPLs.

Next, we present an application of the Social Technology (ST) in two Local Productive Arrangements (LPAs), located in the State of Ceará (Brazil).

7.4 Application of the Social Technology: A Hammock LPA in Jaguaruana/ Ceará and a Crochet[5] Local Productive Arrangement in Nova Russas/Ceará

In this section, using as a reference the morphological and technical frameworks proposed by Bruyne, et al. (1991) for methodological practice, we apply the previously described social technology in two Local Productive Arrangement (LPAs), located in the state of Ceará, Northeast Brazil. The areas of specialization of the cases, both involving relative diversification and a strong local tradition, are: hammocks, located in the municipality of Jaguaruana (CE) and handcrafted crochet, located in the municipality of Nova Russas (CE).[6]

In studying these two cases, we performed on-site exploratory research, including interviews with key local actors, observation and participation (during meetings and informal conversations with the LPAs' participants), as well as the compilation of secondary data through documental analysis. The research was undertaken in the year 2003.

By compiling and analyzing this data we were able to identify participants active in the arrangement under study. The situation of each arrangement was also previously diagnosed, suggesting the dynamics of participative mobilization as a possibility, developing relationships, as the regions institutional, productive and community agents become involved. Based on the proposed Social Technology (ST) for Mobilization of the LPA's, these dynamics suggest the development of a network of interrelationships among the participants, with the potential to generate cooperative actions, to advance in an innovative way on behalf of the local productive arrangement. (Messner and Meyer-Stamer, 2000; Albagli and Maciel, 2003).

Following is a description of the characteristics and proposals for the mobilization of the two LPAs under study, possibly by means of the organizational entities "Forum for Change", "Innovation Laboratories" and "Hearing Points", foreseen in the Social Technology.

---

5   Crochet is a common handcraft in Brazil and may be found in different forms (e.g., garments, linens).

6   Population (2005) of these municipalities are approximately: Jaguaruana equivalent to 32,000 inhabitants and Nova Russas with 30,000 inhabitants. GDPs (2003) are roughly US$38,000,000 and US$28,300,000 and per capita GDPs are US$1,351.1 and US$2,010, respectively. In the year 2000, 70% of Nova Russas population was living in urban areas, whereas in Jaguaruana the correspondent figure was 56%.

7.4.1 Hammock Local Productive Arrangement of Jaguaruana/CE

Hammock production in Jaguaruana has dominated local economy for more than a century and draws back to the times when the state of Ceará was a major producer of cotton in Brazil.[7] With an abundance of raw material, the Jaguaruana Indians ("Black Jaguar", in Tupi language) who lived in the region began to make hammocks for their own use. The production of hammocks, therefore, is part of the local culture and traditions and is characteristically a family undertaking. One cannot ignore the business when reaching the municipality as business invades homes, sidewalks and streets. The major market is the state capital, Fortaleza, located 170 km north of the municipality.

Jaguaruana's hammock LPA currently involves approximately 200 businesses which directly employ about 1,900 people. More recently, a major Brazilian textile group (Santista) established a cotton processing plant in the area which facilitated access to thread by small producers, expanding hammock production even further.

In the last fifteen years, hammock manufacturers of Jaguaruana have faced difficulty in marketing their products, due to competition from São Bento's LPA (neighboring state of Paraíba), which offers lower quality products at a lower price. Although Jaguaruana's hammock has great export potential, production has undergone periods of decline and the small manufacturers' individualism and lack of trust in one another prevented cooperation in overcoming their difficulties. European markets have showed significant interest in Jaguaruana´s hammock, but difficulties faced by local manufacturers, ranging from acquisition of raw materials (lack of working capital, lack of credit, etc.) to shipping the finished product (freight charges, invoices, etc.), have prevented export growth.

Due to these difficulties, in 2002, actors of Jaguaruana's LPA decided to organize themselves and engage in collective efforts. With the assistance of local institutions (e.g., SEBRAE-National Agency for Micro and Small Firm Support) they began a new development phase. Their strategy involved promotion of local public and private governance, including constructions of alliances and cooperation with key institutions and partners. The kick-off event was the creation

---

7   From the 1800s to the 1970s, cotton production was the major cash crop in Ceara. In the 1970s, the crop covered more than 1.3 million hectares in the state. Cotton was then known as the "white gold". It built personal fortunes, brought growth to the state´s countryside, reinforced power of landowners and initiated industrialization (cotton mills) in the state. In the late 1970s, the boll weevil devastated the crop in all Brazil´s Northeast, bringing unemployment to rural areas, decrease of land prices, shutdowns of cotton mills and shift of political power from rural elites to urban-based groups linked to industrial activities. During this period, the cotton-planted area in the state was reduced to less than 1%, compared to during booming times. Textiles industry, however, remained strong in the state, thanks to imports, advanced technology and shift to modern products (denim).

of the Association of Hammock Manufacturers of Jaguaruana (ASFARJA) made up of 27 associates.

In what follows, we explain how the social technology proposed here could be applied by the Jaguaruana hammock arrangement.

Proposal for the mobilization of Jaguaruana's Hammock LPA: Instituting
the "Forum for Change", "Innovation Laboratories" and "Hearing Points"
Assuming that the model of governance based on networks, established through negotiation, is crucial for developing countries, it is essential that an effort be made, by the LPA's participating members to construct a network of interrelationships and dynamic cooperation (Messner and Meyer-Stamer, 2000; Putnam, 2000; Albagli and Maciel, 2003; Beretta and Curini, 2003).

This network can be put into operation through the three organizational entities "Forum for Change", its "Laboratories" and respective "Hearing Points", thereby setting off the processes of cooperative learning and influencing the development of social innovation (Dagnino and Gomes, 2000). Based on the previously cited concepts and assumptions, we present a brief description for the format of these entities in the LPA under observation.

a) Forum for Change   In the case of the Jaguaruana LPA, the Forum for Change can be formed by the following institutions present in the area (Table 7.4).

Table 7.4   Forum for Change (Jaguaruana's Hammock LPA)

Forum for Change
Manufacturers Association (ASFARJA);
Non-Associated Manufactures;
Local government authorities (department of business development);
Local office of SEBRAE-CE (National Agency for Micro and Small Firm Support);
Local representative of Bank of the Northeast (federal development bank);
Local Retailers Association (CDL);
Representative of the State Tax Collection Department (SEFAZ-CE);
Postal Service;
Center for Technological Studies (CENTEC);
SENAI (National Industrial Training Service);
School of Style and Design (Federal University of Ceará – UFC);
Thread Manufacturers Association.

Source: The Authors

b) Innovation Laboratories   The proposals formulated and prioritized by the forum will be handled by the different Innovation Laboratories, established by the Forum´s participating members. The Laboratories are linked to specific problems, chosen by the Forum based on their importance. Suggestions for some of these laboratories are below, as well as their objectives and components.

Innovation Laboratory of Skills Development

This Laboratory focuses on the development of new abilities, both for individuals (human capital), groups (cohesion, and group articulation), and organizations (organizational development) and the LPA´s businesses.

From the start, the creation and functioning of the Forum, Innovation Laboratories and Hearing Points will require that the participants be prepared to work in groups. It is therefore recommended that the LPA's participants be offered courses and training on the basics of group participation, collective organization, association and cooperation (Table 7.5).

Table 7.5   Innovation Laboratory of Skills Development – Examples of Actions (Jaguaruana's Hammock LPA)

Examples of Actions
Interpersonal relationships;
Group dynamics (aimed at stimulating interaction and trust among the LPA´s members);
Basic topics on associations and cooperatives;
Safety and hygiene in the workplace;
Labor laws;
Development of business networks.

Source: The Authors

Innovation Laboratory for Technology and Design

This purpose of this Laboratory is to research trends, styles and new opportunities for diversification, increase in value-added and quality enhancement of LPA's products (e.g., use of new dyes, raw materials, equipment). An example of this laboratory's possible contribution would be the recruitment of professionals to teach courses and offer consulting services to manufacturers. Another example would be the establishment of partnerships with training institutions, such as the National Industrial Training Service-SENAI and the School of Style and Design of the Federal University of Ceará (UFC), with the purpose of perfecting the design, style and quality of the products manufactured in the LPA, as well as diversification of items (Table 7.6).

Table 7.6   Innovation Laboratory for Technology and Design – Examples of Actions (Jaguaruana's Hammock LPA)

Examples of Actions
Courses and consulting services to improve the quality of products and production processes;
Development of a seal of quality and/or a certificate of origin for the LPA´s products;
Agreements with institutions such as universities, technology agencies, research centers with the following objectives: - Analyze environmental aspects (dyeing: use of natural dyes, less polluting ingredients, more efficient and environmental-friendly processes); - Enhance use of naturally colored cotton for the production of threads, hammocks and other items produced in the LPA.
Partnerships with universities and technology schools (e.g., University of the Vale do Acaraú-UVA and CENTEC), research centers (National Agriculture Research Agency-EMBRAPA and rural extension agencies (Ceará Rural Extension Agency-EMATERCE) aiming at: - Improvement of design, colors and use of inputs; - Equipment maintenance, adaptation and improvement of machines; - Improve energy use efficiency, in order to allow for lower production costs.

Source: The Authors

Innovation Laboratory for Finances

This Laboratory focuses on discussing problems relating to value-added tax (ICMS) relative to thread and hammocks in the state; tax collection procedures, such as issuing of invoices, in-transit items and other related subjects; as well as topics related to costs and price formation. Key agencies involved would be the State Tax Collection Department (SEFAZ), State Department of Industrial Development, municipal government authorities and SME Supporting Agency — SEBRAE (Table 7.7).

Table 7.7   Innovation Laboratory for Finances – Examples of Actions (Jaguaruana's Hammock LPA)

Examples of Actions
Creation of a working group to discuss and propose solutions for LPA's tax related problems;
Workshop on price structure, efficient cost control mechanisms and pricing methods;
Monitoring of the LPA's sales and trends;
Discussion on strategies to face defaulting (e.g., creation of collective client directory with information on high-risk clients);
Discussion on alternative credit lines for including working capital and investment.

Source: The Authors

Innovation Laboratory for Communication, Marketing and Market Expansion

This Laboratory focuses on devising strategies to open new markets for the LPA, in addition to developing a trademark for the local products and encouraging new sales initiatives/techniques (Table 7.8).

Table 7.8   Innovation Laboratory for Communication and Marketing (Jaguaruana's Hammock LPA)

Examples of Actions
Create a catalog featuring the products of the LPA
Create a website for the LPA, including tools to allow for direct sales to consumers
Establish a showroom to expose Jaguaruana's hammocks in a strategic location
Negotiate special courier and freight rates for the Jaguaruana's products.

Source: The Authors

c) Hearing Points   Based on data obtained from observation, contacts and interviews, we propose that the Hearing Points for Jaguaruana's LPA include the following institutions and organizations: School of Style and Fashion Design, School of Agriculture and Chemical Engineering School, Schools of Business Administration, International Trade, Public Policies; the Brazilian Enterprise for Agricultural Research — EMBRAPA (department of cotton research); Textile Manufacturers Association; Ceará Federation of Industry-FIEC; Center for International Information-CIN; Bank of the Northeast—BNB; suppliers of textile products, export agents, as well as representatives of foreign countries in Ceará (e.g., Germany, which has shown interest in Jaguaruana's hammocks).

7.4.2 Crochet Local Productive Arrangement of Nova Russas/CE

Located in the heart of the Ceará semi-arid region, approximately 300 km southwest of Fortaleza, Nova Russas claims to be the "Brazilian capital of crochet". It is sure a fair statement. The municipal economy is driven by crochet activities: artisans, apparel firms, suppliers of inputs, market intermediaries, wholesalers are easily seen anywhere at almost anytime in the municipality.[8] There are approximately 10,000 crochet artisans (men and women of all ages) working in Nova Russas and its vicinities. Informal estimative indicates that the income generated by this

---

8   One of the most important commercial events happens daily in front of the main Catholic church, before the sunrise, from 2am to 6am (event is known as the "Owl trade fair"). Initial reason behind this time choice was to prevent presence of tax agents and consequent detection of tax evasions.

activity frequently exceeds municipal total revenue, including federal revenue transfers to the municipality (roughly equivalent to US$ 8 million per year), which is mostly unusual for small Northeastern municipalities, such as Nova Russas.[9]

Proposal for the Mobilization of Nova Russas' LPA: Establishing the "Forum for Change", "Innovation Laboratories" and "Hearing Points"
Nova Russa's LPA has an association (Association of the Crochet Artisans of Nova Russas — ASCRON), but only 1% of the region's artisans are members of this institution. ASCRON is in charge of marketing products of its members and cannot afford to enlarge unlimitedly membership because it does not have capacity to substantially expand sales. Majority of artisans are left isolated and unorganized, hence facing innumerous problems, ranging from marketing of products to lack of training. Increasing the level of participation and organization of artisans and others involved in the local crochet arrangement is key to promote growth, innovation and arrangement development (Putnam, 2000; Albagli and Maciel, 2003; Beretta and Curini, 2003). This process aims at strengthening the cooperative bonds between the productive participants of the LPA and local public authorities, as well as other members of society (community representatives, research and credit institutions), by means of periodic meetings and other activities, such as lectures and workshops targeting subjects of common interest (Rullani, 2003; Dagnino and Gomes, 2000). The process can be activated by the technology proposed here which focuses on strengthening social mobilization of the arrangements participants, through intensification of relationships and promotion of systematic practices through network development (Messner and Meyer-Stamer, 2000; Cassiolato and Lastres, 2003).

a) Forum for Change   This Forum should include participation of representatives of all sectors and agencies involved in the local crochet business. We have identified the following institutions as key members of this LPA´s Forum (Table 7.9).

b) Innovation Laboratories   Given its lack of marketing skills, ASCRON and other arrangement producers would benefit from effort to promote sale expansion and opening of new marketing channels. The arrangement also needs to increase its level of visibility to potential consumers, an accomplishment that would benefit the municipality as whole, as it would bring new income and reinforce municipal image as a specialized territory. In general, there is a lack of information about this arrangement in the state, a fact that limits the possibility of placing the products in more lucrative markets outside the municipality. Furthermore, it is important to

---

9   In Brazil's Northeast, most small municipalities survive thanks to resources transferred by the federal government. These include percentages of the FPM (Municipal Participatory Transfer Fund), ICMS (state-collected sales tax) and other funds. Revenues directly collected by the municipalities (e.g., property tax, service tax) are usually an insignificant share of total municipal revenues.

Table 7.9   Forum for Change (Nova Russas' LPA)

Forum for Change
Crochet Artisans Association (ASCRON)
Local office of SEBRAE (National Agency for Micro and Small Firm Support);
Representative of Municipal Government (Department of Economic Development and the Environment – SEDEMA)
Representative of the State Government (State Tax Collection Department-SEFAZ, Handcrafts Development and Commercial Center of Ceará-CEART)
Local representative of Bank of Brazil
Local representative of Bank of Northeast of Brazil (BNB)
Federal University of Ceará/School of Style and Fashion
Dragon Fashion (Organization of Professional Designers of Ceará)
Representative of the local Retailers Association – CDL
Buyers and suppliers
Representatives from NGOs operating in the region and focusing on small businesses and local development

Source: The Authors

improve the product's design and upgrade the artisan's skill level, particularly in respect to use of colors and patterns. The three following laboratories are suggested for the arrangement.

Communication, Marketing and Market Expansion Laboratory
In order to provide greater visibility for the products of this LPA, the focus should go to visual communications strategies, with emphasis on marketing, such as: elaboration and implementation of marketing plans, advertising of products in different media, including the internet, and dissemination of information on local, national and international competitors within arrangement members (Table 7.10).

Technology and Design Laboratory
This Laboratory aims to research trends, styles and new opportunities for diversifying and increasing the value of the LPA's products, as well as improving the quality of the products (color combinations, patterns, new materials, etc.). As an example of a possible action, the laboratory might choose and hire specialized professionals to minister courses and provide consulting services on those issues. The Laboratory could promote partnerships with training agencies (e.g., SENAI and the School of Fashion Design, located in Fortaleza) in order to improve product design and styles

Table 7.10   Innovation Laboratory for Communication and Marketing (Nova Russas' LPA)

Examples of Actions
Placement of visual signs (downtown area, major roads crossing the municipality and surroundings) to indicate location of arrangement and its commercial poles
Publicize arrangement products in key locations or websites (Fortaleza's airport, hotels, Fortaleza'ss tourist sites, SEBRAE Commercial spaces, CEART Stores, Central Market, EMCETUR (Handcraft and Tourism Center), Fortaleza's Bus Station
Preparation and distribution of product catalogs, folders and other promotional material
Use of advertisement targeting potential buyers (e.g., direct mailing)
Insertion of articles and information about arrangement products in specialized fashion magazines
Create a website for the arrangement
Participation in regional and national trade fairs
Establishment of a showroom in highly visible location
Elaboration of a Buyer's Directory
Development of a trademark (inspired by the municipal history and culture) and a seal of quality for the products
Further diversification of products (e.g., use of crochet in various types of apparel, such as beachwear, jeans, purses, etc.)
Negotiation of special freight rates and payment terms with freight companies, postal and courier services

Source: The Authors

Table 7.11   Innovation Laboratory for Technology and Design (Nova Russas' LPA)

Examples of Actions
Collaborate with the Ceará Design Center, SENAI, UFC's Fashion Design course and Dragon Fashion, in the following: design, patterns, seasonal collections, color combinations, forms and new materials, financing for consulting services and course conclusion assignments, such as notions of chromatography.
Facilitate the development of information channels among: -   Artisans, retailers and sewing thread manufacturers (color catalogs, samples of new products, indication of trends); -   Artisans and buyers (suggestions for new prints, sizes, etc.).

Source: The Authors

Laboratory for Skill Development
The focus is on the development of new skills, as much for people (human resources) and groups (cohesion and group articulation) as for organizations (organizational development), and is comprised of the businesses and other institutions present in the LPA's area (Table 7.12).

Table 7.12   Laboratory for Skill Development (Nova Russas' LPA)

Examples of Actions
Social Development: performance of group facilitation and interpersonal development exercises, group dynamics aimed to foster trust and promote group cohesion
Technical Courses in Sales and Exporting, including legislation, notions on how to establish an export cooperative and technical exporting procedures
Promotion of entrepreneurship (motivation, business vision), training on associations and cooperatives
Basic training on cost formation, covering minimal notions of cost items, cost calculation and pricing for the LPA's products
Literacy courses, as a number of artisans cannot read, interpret or write basic Portuguese language

Source: The Authors

c) Hearing Points   Various institutions can act as Hearing Points for this LPA, including the Dragon Fashion Organization, SENAI, and specialized websites (in Brazil, examples include: *São Paulo Fashion Week, Rio Fashion Week, Textília,* Anhembi University etc.), specialized magazines, regional and national apparel trade fairs, fashion shows, schools of Style and Design, export agencies/brokers, textile technological centers and thread manufacturers. The key objective here is to acquire new and external ideas and information, which could be used by arrangement actors to promote development and strengthening of the LPA (Maggioni and Riggi, 2002).

7.5 Final Considerations

The study of productive agglomerations, focusing on arrangement and local productive systems, offers a comprehensive understanding of the operational dynamics of a region's SMEs, as it demonstrates how the competitiveness of a productive agglomeration is affected by factors such as social capital, governance and cooperative practices.

In this essay, we have illustrated, through two empirical cases, the application of a Social Technology for mobilization of local productive arrangements. The Social Technology presented here is based on the concepts of social capital (Putnam,

1999; Albagli and Maciel, 2003; Beretta, Curini, 2003) and network governance (Messner and Meyer-Stamer, 2000), and entails the development of three types of organizational entities, structured as networks, and aiming to contribute to social innovation (Dagnino and Gomes, 2000) through strengthening of LPA's social capital and governance. Final expected outcome corresponds to arrangement growth and local economic development.

Social Technology (ST), considered as both a theoretical and practical challenge, was applied to the two LPAs under observation. Initial effort of this study included arrangement mapping, with the purpose of identifying major arrangement actors and possible relationships among them. Each arrangement was also previously analyzed and, as a subsequent step, authors have proposed the above approach to mobilize arrangements productive agents, institutions and community, through the development of a network of interrelationships.

The Social Technology proposed here suggests the establishment of three types of organizations, set up as networks and based on cooperative learning. Initially, a larger group of selected participants is identified ("Forum for Change"), which will then subdivide into several working groups ("Innovation Laboratories"), focused on specific assignments considered to be essential in solving the LPA's problems. Additionally, these groups are explicitly linked to institutions that can serve as a reference ("Hearing Points") for the arrangement. Basically, the hearing points should encompass the latest know-how or vision related to subjects of importance to the LPA.

The reason for establishing the "Forum for Change" is to develop a network of relationships for the arrangements participants, intensifying interaction among them and encouraging systematic and organized collective actions, thereby catalyzing the atmosphere of cooperative learning and strengthening the LPA's social capital. A systematic attempt to solve the LPA's problems requires coordination, hence likely to promote governance. These two assets, social capital and governance, are essential ingredients for the LPA's strengthening and growth, given the small scale of majority of their businesses.

Finally, this essay's intended contribution is to develop and disseminate a Social Technology for productive arrangement mobilization that has the potential of strengthening LPA's, by increasing the quantity and intensity of their links with other major players in the territory, in order to consolidate their support to arrangement in reference. The function of the "Forum for Change", the "Innovation Laboratories" and the "Hearing Points" is to unify efforts to identify opportunities and resources of various types, define a strategy to enhance arrangement competitiveness and implement it effectively.

A next step in the development of this approach will be the actual implementation of this social technology, in order to test its feasibility and effectiveness. This would be possible through a social pact promoted by the arrangement's governance and the local government, with the aim at developing the territory.

The final result of this theoretical-methodological effort might be to assist policy makers in formulating development policies in these territories, with the objective of strengthening the agglomerations of small- and medium-sized businesses, and creating the means for continuous evaluation, which will permit the necessary adjustments as they become necessary.

References

Albagli, S., Brito, J. (2002). Arranjos Produtivos Locais: Uma nova estratégia de ação para o SEBRAE — Glossário de Arranjos Produtivos Locais. Rede Sist.

Albagli, S., Maciel,M.L. (2003). Capital social e empreendedorismo local. Proposição de política para a promoção de sistemas produtivos locais de micro, pequenas e médias empresas. 2003, www.ie.ufrj.br/redesist <http://www.ie.ufrj.br/redesist>. Accessed on 18.06.2006.

Amaral Filho, J. Do (1999). A endogeinização no desenvolvimento econômico regional, Annals of the ANPEC, XXVII Encontro Nacional da ANPEC, Belém-Pará, Brazil, December,. 1281–1300.

Amorim, M.A. (1998). Clusters como estratégia de desenvolvimento industrial no Ceará. Fortaleza: Banco do Nordeste, ETENE.

Amorim, M.A., Moreira, M.V. and Ipiranga, A.S. R. (2004). Constructing governance within small firm cluster: A view from the developing world". Annals of the BELL CONFERENCE IIT: Building a sustainable city through sustainable enterprise, Stuart Graduate School of Illinois, Chicago, USA.

Audrtsch, D. B. (1998). Agglomeration and the location of innovative activity. *Oxford Review of Economic Policy,* 14(2).

Baptista, Creomar (2003). Distritos flexiveis e desenvolvimento endógeno: Uma abordagem "marshalliana" <http://www.geocities.com/statprof/distrit.html> . Accessed on 20.01.2003.

Britto, J. and Albuquerque, E. da M. (2003). Caracteristicas estruturais de Clusters Industriais na economia brasileira: uma análise inter-setorial. http://www.nepp.unicamp.br/cadernos/cadernos/caderno39.pdf. Accessed on 22. 01.2003.

Beretta, S. and Curini, L. (2003). Il ruolo della famiglia nel generare capitale sociale: um approccio di economia politica. In *VIII Rapporto CISF sulla famiglia in Itália* , 290–339. Milano: S. Paolo.

Blessi, G. T. (2002). Creative milieu e competitività: um nuovo paradigma di sviluppo locale. In Osservatorio Impresa e Cultura, *Cultura e competitività. Per un nuvo agire imprenditoriale*. Roma: Rubbettino Editore.

Bourdieu, P. (1985). The forms of capital. In J.G. Richardson (ed.), *Handbook of Theory and Research for Sociology of Education* , 241–58. New York: Greenwood.

Brusco, S. (1990). The idea of the industrial districts: its genesis. In Frank Pyke, Giacomo Becattini and Werner Sengenberger (eds), *Industrial Districts and Inter-firm Cooperation in Italy*. Geneva: International Institute for Labour Studies/ILO.

Brusco, S.; Solinas, G. (1999). Partecipazione necessaria e partecipazione possibile. *L'impresa al plurale. Quaderni della partecipazione*, 3–4, 411–28.

Bruyne, P.; Herman, J. and Schoutheete, M. (1991). *Dinâmica da Pesquisa em Ciências Sociais. Os Pólos da Prática Metodológica.* Rio de Janeiro: Editora F.Alves.

Castells, M. (1986). Mudança tecnológica: reestruturação econômica e a nova divisão espacial do trabalho. *Espaço e Debates*, 17, ano II.

Cassiolato, J. E. and Lastres, H. M.M. (2003). O foco em arranjos produtivos e inovativos locais de micro e pequenas empresas. In H. M. M. Lastres, J.E. Cassiolato and M. L. Maciel, *Pequena empresa: cooperação e desenvolvimento local.* Rio de Janeiro: Relume Dumará.

Cassiolato, J.E. and Szapiro, Marina (2002). Proposição de políticas para a promoção de sistemas produtivos locais de micro, pequenas e médias empresas- Arranjos e sistemas produtivos locais no Brasil, Instituto de Economia da Universidade Federal do Rio de Janeiro- IE/UFRJ, Rio de Janeiro.

Coleman, James S. (1988). Social Capital in the creation of Human Capital. *American Journal of Sociology*, volume 94 Supplement, 95–120.

Dagnino, R. and Gomes, E. (2000). Sistema de inovação social para prefeituras. Annals of the CONFERÊNCIA NACIONAL DE CIÊNCIA E TECNOLOGIA PARA INOVAÇÃO. São Paulo.

Fukuyama, F. (1996). *Confiança. As virtudes sociais e a criação da prosperidade.* Rio de Janeiro: Rocco.

Humphrey, J. and Schmitz, H. (2000). Governance and upgrading: linking industrial cluster and global value chain research. IDS Working Paper 120. Institute of Development Studies, UK: University of Sussex.

Ipiranga, A. S. R. (2006). Os arranjos e sistemas produtivos locais entre aprendizagem, inovação e cultura. Annals of the XXX ENCONTRO DA ASSOCIAÇÃO NACIONAL DOS PROGRAMAS DE PÓS-GRADUAÇÃO — ENANPAD, Salvador.

Krugman, P. (1993). *Geography and Trade.* Cambridge, USA: MIT Press.

Lalkaka, R. (1997). *Supporting the Start and Growth of New Enterprises.* NewYork : United Nations Development Programme.

Le Borgne, D. (1991). *La politique industrielle regionale en Italie.* Paris : Ministère de l'industrie et de l'Aménagement du Territoire/CEPREMAP.

Marshall, A. (1996*). Princípios de Economia.* São Paulo: Nova Cultural.

MCT — Ministério da Ciência e Tecnologia (2001). Manual para implantação de incubadoras. <www.mct.gov.br/setec.htm <http://www.mct.gov.br/setec.htm>. Accessed on 06.11.2001.

Maggioni, M.A., Ricci, M. (2002). Forme alternative di collective learning: um approccio sistemico-popolazionista ed alcune simulazioni. In R. Camagni and R. Capello, *Apprendimento colletivo e competitività territoriale.* Milano: Franco Angeli.

Messner, D. and Meyer-Stamer, J. (2000). Governance and Networks. Tools to study the dynamics of clusters and global value chains. Paper prepared for the

IDS/INEF Project The Impact of Global and Local Governance on Industrial Upgrading. University of Duisburg.

Piore, M. and Sabel, C. (1984). *The Second Industrial Divide: Possibilities for Prosperity.* New York: Basic Books.

Porter, Michael E. (1998). Clusters and the new economics of competition. *Harvard Business Review*, 76(6), 77–90.

Putnam, R. (1993). *Making Democracy Work: Civic Traditions in Modern Italy.* Princeton: Princeton University Press.

Pyke, F., Becatitini, G. and Sengenberger, W. (eds) (1992). *Industrial Districts and Inter-Firm Co-Operation in Italy.* Geneva: International Institute for Labour Studies.

Secretaria Executiva da Rede de Tecnologia Social (2004). Tecnologia social: uma estratégia para o desenvolvimento. Rio de Janeiro: Fundação Banco do Brasil.

Rullani, E. (2003). Intelligenza terziaria e reti professionali il nuovo motore dello svluppo. In: Il terziario motore di sviluppo dell´economia: il side per il management. Milan: FENDAC.

Saxenian, A. (1994). *Regional Advantage: Culture and Competition in the Silicon Valley and Route 128*. Cambridge, MA: Harvard University Press.

Schmitz, H. (1994). *Collective Efficiency: Growth Path for Small-Scale Industry*. Brighton: IDS.

Schmitz, H. and Nadvi, Khalid (1999). Clustering and Industrialization: Introduction. *World Development,* 27(9), 1503–1514.

Tendler, J. (2002). Small Firms: the informal sector and the devil´s deal. *IDS Bulletin*, 33(3).

Tendler, J. and Amorim, M. A. (1996). Small firms and their helpers: Lessons on demand. *World Development,* 24(3), 407–26.

Chapter 8
Conclusions: Lessons from the Cases

Jose Antonio Puppim de Oliveira[1]

The cases in this book are a good sample to draw some conclusions on how clusters of small firms in developing countries can innovate and socially upgrade. Even though they do not represent an extensive and comprehensive sample, the cases present an interesting diversity of sectors, contexts and actors involved.

Clusters of small firms (SMEs) in developing countries face tremendous challenges to socially upgrade. These clusters and their firms generally lack human, financial and technical capacity to innovate and improve their economic/social/labor/environmental standards, and consequently the well being of the communities they are located. There is still no general framework to describe why and how clusters socially upgrade, but this book provides some analyses to understand how some clusters were able to move from a group of backward producers lacking compliance with the minimum standards of formalization, labor and environment quality to a more dynamic, innovative, formal, law-compliant and environmentally-friendly agglomeration. They do not represent ideal "perfect" cases of social upgrading, but they serve to understand a bit more the processes and conditions under which clusters socially upgrade, and advance the research agenda on the topic.

The conclusions are organized to answer the three questions posed in the introduction (Chapter 1). I tried to use the lessons from the cases to develop the arguments that give one answer to the each of the questions. It is not expected that they offer definitive comprehensive responses to the complex issues addressed in the questions, which would need extensive structured research efforts, but they provide some directions for pursuing further research in the future.

How and Why Were Clusters Able to Socially Upgrade Without Losing Competitiveness?

Many of the examples in this book are clusters that were able to socially upgrade and also gain competitiveness through the traditional (economic/production)

1 Development Planning Unit (DPU), University College London (UCL), UK, and Laboratorio do Territorio (LaboraTe), the University of Santiago de Compostela (Spain). During the edition of this book he was at the Brazilian School of Public and Business Administration (EBAPE) of the Getulio Vargas Foundation (FGV), Rio de Janeiro (Brazil).

upgrading. Indeed, those issues (social upgrading, economic upgrading and competitiveness) are directly linked in many cases. Many firms and clusters that socially upgraded could become more efficient, develop new products or have the opportunity to access premium markets. For example, the producers of organic coffee in Chiapas, Mexico, could sell their coffee in European markets, which paid a premium price for their products (see Damiani's chapter in this book). This allowed them to develop a quality coffee and use sustainable production methods.

Sometimes the gain in competitiveness is a necessary condition to allow firms or clusters to socially upgrade. The development of new socially responsible products, formalization of workers or the use of a more sustainable input can be difficult in practice if these upgradings do not lead to more efficient processes or the markets do not reward the changes. The case of the furniture industry in Indonesia (Posthuma's chapter) showed the constraints for using sustainable timber as an input (instead of regular timber, in general from unsustainable logging). Firms did not want to use the more expensive certified sustainable timber because this would increase the costs of inputs and the exporting market would not differentiate the prices. In a globalized economy, if exporting firms insisted in using sustainable timber, they could become economically unviable and consequently lose clients or even close down. Thus, firms prefer regular timber, even though this source can be unsustainable and more expensive in the long run, as timber supplies become more scarce and distant. Indonesian firms continue to produce because there is no other alternative in the short term.

The social upgrading in the Indonesia case would possibly be economically viable if the use of sustainable timber would come together with the improvement of the quality of the product (upgrading the product). If firms were able to find premium markets for their furniture using, for example, traditional carving techniques in parts of the furniture, they would be able to pay for the sustainable timber and still be competitive. On the other hand, they would move to a more sophisticated market and also have more sophisticated competitors, such as Italian producers.

However, gains in competitiveness with traditional (production) upgrading is not a sufficient condition for spurring social upgrading automaticaly. Increase in competitiveness can lead to a situation of augment in production in certain clusters with social, labor and environmental consequences going unregulated. In many cases, external actors, which are not part of the clusters, play an important role in reversing this situation and making clusters socially upgrade. Political and social action are also important to adjust the economic gains of upgrading, or even making it to happen in the long run, allowing both social and economic upgrading. For example, the development of relatively good quality jeans with low prices led to an increase in water consumption and pollution in a water-scarce region of Toritama in Brazil (see Almeida's chapter). This situation would have continued if the new public prosecutor did not act and apply pressure for the enforcement of environmental legislation to both the firms and enforcerment agencies. Firms had to comply with the law, and actually some firms had to close down. However, in the log run, by investing in water recycling technology, firms could save water (an expensive input) and become competitive in the long run. Thus, the pressure from an external actor (public prosecutor) outside the market was fundamental for social upgrading.

Upgrading (both production and social) generally involves innovation capacity. Firms and clusters have to innovate to improve social, labor and environmental standards and become competitive. The need for innovation and the innovation capacity of firms depend much on the sector and markets they operate. Firms in some sectors, generally more technologically intensive such as the Information Technology (IT), are required to innovate constantly to be in the market. The innovation capacity depends on constant training of the workforce and investments in Research and Development (R&D). For example, the Indian software industry requires the formation of high skilled workers even for small firms (see Okada's chapter). They need such a workforce to be able to keep the high standards required for the markets in which they are immersed. Indian software small firms need to hire some of the best engineers available, sometimes competing with the salaries of Multinational Corporations (MNCs), in order to keep pace with their clients' requirements (especially Americans) and be competitive globally. The workforce requirements and high demands in the labor market make it difficult to have informal workers and bad working conditions. Many times, small firms have to invest substantially in R&D, which can be important for upgrading and being competitive. This innovation capacity also makes small firms competitive in certain markets even competing with larger firms. Firms with such capacity can innovate and adapt to any requirements, be it technological, social, environmental or labor.

The governance within clusters is another important factor to explain social upgrading. Producers and their supporters have to organize themselves to innovate and improve social, labor and environmental standards. The way they are able to do that depends on some external and internal dynamics, such as markets, production processes, relationship among members of the cluster and organizational capacity of the firms and their supporters. The existence of an organization of producers (e.g., association, union or cooperative) is important for negotiating with buyers, representing the interests of the producers economically and politically and organizing them to complying with standards. In the cases of producers of tobacco in China and coffee in Mexico (see Damiani's chapter), producers had to comply with contracts and social and environmental quality standards. Their associations had to organize production, train farmers and monitor producers to ensure that the goods had good quality standards, such as fit organic methods of production required by clients abroad in the case of Mexico (Damiani's chapter).

Supporting organizations also have a crucial role in the governance of clusters for social upgrading. They provide information about markets, training, technical expertise and even be prepared to certify products and processes. In the case of jeans production in Toritama (see Almeida's chapter), supporting organizations were key to develop a customized technology for industrial sewage treatment in order to solve the water pollution problems denounced by the public attorney. They were able to mobilize experts locally and internationally, search for local and external funding for innovation and install the equipments in the firms with their consent. In the furniture clusters in southern Brazil (see the chapter by Puppim de Oliveira), a supporting organization (SENAI-CETEMO, Center of Furniture Technology) was fundamental for providing training and information on environmental and social demands of external markets and society, as well as qualifying itself to make tests and certify several quality requirements, including environmental standards.

The efforts of the cluster-supporting organizations can be amplified by efforts of mobilization of the different actors within and outside the cluster. Actors need to interact with each other to create social capital for developing trust among them and better governance mechanisms to facilitate cooperation and joint action. The social technology for mobilization described in the chapter by Ipiranga, Faria and Amorim is an initiative to try to improve the governance of the clusters by creating "channels" of interaction among firms, clusters members and outside organizations and individuals. Those channels would allow cluster members to identify new market and technology trends, as well as new demands in terms of environmental and social standards.

What were the Roles of External Actors, such as Governments, NGOs and Clients in Cluster Upgrades?

As I discussed earlier, external actors play fundamental roles in social upgrading in all three frameworks. They generally are those who "break" the clusters' dynamics by pressing for law enforcement, bringing new ideas and technologies, providing information, and creating and making market demands on social and environmental standards.

There is a myriad of external actors in the case studies in this book, national and foreign, ranging from international clients to public attorney. They can be market-related and non-market actors. Market-related external actors are individuals and organizations that are not direct parts of the cluster but influence the cluster through the chain, such as final clients and suppliers. For example, clients can make demands on their suppliers for social, labor or environmental standards. In the case of the furniture clusters in Brazil (Puppim de Oliveira's chapter), foreign clients, especially the Europeans, asked for different kinds of environmental certifications such as ISO 14001 (environmental management system) or FSC (timber certification). Many times they were not the direct clients of the firms in the cluster, but their demand came through the local export agents. Market actors can also help the firms to innovate and upgrade by providing consultants, information or financial resources to firms in order to adapt to the market demands. Indians residing abroad are one of the main sources of market and technological information for IT small firms in India (see Okada's chapter). In Mexico, coffee producers were able to learn through their clients about opportunities in the market of organic coffee and contact certification organizations (see Damiani's chapter).

Non-market external actors affect change in firms and clusters through political and social pressure. Those actors could be public attorneys, NGOs, government officials or even the press. The political and social action, as well as new ethical concerns, and not only market-driven economic causes, are important to drive social upgrading. Non-market external actors are key to make firms comply with certain environmental, labor or social regulations, as well as provide information, training or technical assistance. For example, social and labor organizations are some of the main sources of pressure for firms' compliance with local and international labor standards. In many developing countries, those organizations are weak or inexistent, making labor suffer from widespread informality, child or slave labor and low salaries. The high informality rate in the labor market of the furniture cluster in Indonesia can be a consequence of the weak role of labor unions (see Posthuma's chapter). The case of the jeans cluster in Brazil (Almeida's chapter) illustrates how the public attorney made firms to comply with the existing environmental regulation. He catalyzed the role process of change in the relation of the cluster towards the environment.

Some external actors influence the firms through non-market actions but have market consequences. They can provide information, certification, technical assistance or consultancy for clusters or firms to address certain social/labor/ environemntal issues that have markets consequences, such as improve process efficiency or open new market opportunities. For instance, NGOs can help producers to adapt their plantations to organic production (see Damiani's chapter). In Mexico, international NGOs helped the local farmers association to adapt and get certified to international organic standards, thus farmers could sell to premium coffee markets in Europe.

Markets alone are not enough to drive changes in firms and clusters to socially upgrade in a large number of the cases, even premium export markets in developed countries. Other sources of change are necessary to catalyze social upgrading. For example, the access to export markets by the furniture clusters in Indonesia (Posthuma's chapter) did not lead to an increase in formalization of companies or workers, or better environmental practice. However, markets can be an important factor to catalyze change and have some social and environmental impact on firms.Some markets, especially those more demanding in terms of social/labor/ environmental standards, affect mostly the leading firms and clusters—those that are prepared both at the firm and cluster level to innovate and reach the standard demands. Markets can bring new information and provide an incentive to the firms that innovate and upgrade. Thus, both market and non-market organizations are important to catalyze social upgrading of small firms and their clusters.

What were the Key Factors Determining the Effectiveness of the External Interventions?

The effectiveness of the external intervention to make clusters innovate and socially upgrade depends on several factors, especially those related to the cluster context (sector, markets, governance, social and political situation, etc.) and the kind of relation between external actors and cluster actors.

The intervention may vary according to the cluster features. For example, SMEs in clusters in some competitive sectors, such as IT, that depend on a highly qualified workforce and constant innovation do not present many problems related with low salaries and labor informality, as compared to other clusters based on low technology processes and price competition, such as the furniture cluster in Indonesia. The latter may need a much stronger external intervention to formalize labor (e.g., support to labor unions or enforcement organizations) than the former. Thus, policymakers should understand the cluster context in order to design social upgrading policies.

Clusters may have problems regarding some social/labor/environmental aspects, but not others. On the other hand, upgrading policies may be able to tackle certain problems, but not others. For instance, the intervention of the public attorney to combat water pollution in the jeans clusters in Toritama (Almeida's chapter) had positive effects, but the cluster still suffered from bad working conditions and informality (both labor and formalization of companies). Public pressure and the nature of the external actor may deliver the results of the policy.

The kind of relationship between the actors in the clusters and external organizations and individuals are key to drive innovation and social upgrading. Both the market and non-market external actors and interventions need a certain degree of cooperation and trust, and this needs certain time to be built, to develop a relationship that leads to social upgrading. From the cases in this book, the kind of relation between cluster and external actors should be of working together by fostering problem solving and innovation. The pure market relationships (arm's-length in Table 1.1) in the upgrading demands do not help to improve standards locally. If an external client just says "comply with those norms or we will buy from other suppliers" would not solve the local problems in many cases. This may also be expensive for the buyer/client, as it would have to expend time and resources to find other reliable suppliers who comply with the required social/labor/environmental norms. The social upgrading cases driven by markets were those with a network or quasi-hierarchical kind of relationship (described in Figure 1.1). The cases of Mexico and China (see Damiani's chapter) are good examples of social upgrading through relationship between clients and producers in order to solve specific problems of quality. In Mexico, European clients (directly and through NGOs) decided to invest and work together with the association of producers to transform their product in a certified organic coffee.

The quasi-hierarchical interactions between buyers and firms can promote, but also restrict continuous innovation in the cluster or limit social upgrading in specific areas (e.g., environmental standards). SMEs in quasi-hierarchical relation with a leading buyer can receive human and technical resources, as well as incentives, from their buyer to innovate and socially upgrade. However, there is no incentive for firms to go beyond the social/environmental standards required by the client sometimes. For example, the furniture clusters in southern Brazil (Puppim de Oliveira's chapter) the quasi-hierarchical relation between firms and final clients or agents did not give incentives or allow innovations in social/labor/environmental standards that went beyond what was established in the contracts. Firms were expected to reach the established standards only (and not beyond), generally those established by the environmental certifications (ISO14001 or FSC), but there was no incentives to reach higher social/labor/environmental standards.

The participation of the external actors of the public sector appeared in all cases of social upgrading studied in this book. They play different roles. On the one hand, some public organizations have the role to foment firm innovation and social upgrading, such as giving training and information, connecting firms to potential markets or certifying products. For example, in the Brazilian furniture clusters the SENAI-CETEMO was key in upgrading the cluster by training workers and testing the furniture. However, fomenting initiatives do not reach all firms in the cluster. Generally, only the leaders engage and take advantage of those initiatives to innovate and upgrade, but many others stay informal in the bottom, limiting the effect of foment initiatives.

On the other hand, the power of legitimated coercion is the feature that differentiates some actors in the public sector from other external actors. Law enforcement agents have the power to make all and every firm to comply with the social, labor and environmental standards. However, law enforcement is still weak in many parts of developing countries. Public organizations suffer from lack of human, financial and technical resources for proper law enforcement, as well as problems like corruption and lack of motivation.

Accountability is weak on those public organizations from the civil society, especially firm managers or owners, who are some of the leaders in many of poor localities where clusters are located. As many parts of the population in poor areas, many firms are totally or partially informal and do not pay direct taxes, so they cannot press governments politically to improve public organizations and services, such as infrastructure, which are key for allowing upgrading. Also, because firms are informal and generally do not pay taxes, governments do not feel responsible for having to provide adequate services or public policies or help to upgrade them.

Moreover, when some enforcement organizations try to enforce social/labor/ environmental laws or combat informality often they suffer resistance from firms and other actors in the clusters, including government organizations such as local governments. Even though they recognize the lack of law enforcement, those actors see the enforcement of the law as costs and a drawback in their (price) competitiveness. Firms often claim that they would close down if they have to bear the costs of economic/social/labor/environmental regulations. Thus, they try to block the action of law enforcers, sometimes asking the support of politicians making a deal: "do not enforce the law on us and we (cluster members) support you politically" (the devil's deal described by Tendler). These kinds of deals undermine many initiatives to enforce the law, which can lead to an institutional environment of widespread corruption and informality, generalized conflicts and lack of motivation of officials in public organizations.

Public organizations have to take a different approach to deal with clusters of SMEs in order to promote social upgrading effectively. Sometimes, there are public organizations which become involved in a cluster by both working on fomenting social upgrading and enforcing the law, but they do not coordinate the actions. Instead of using only the "foment initiatives" or the "police power" for law enforcement separately, they have to combine those two initiatives by providing support to help firms to upgrade and comply with the law. It would be a kind of "carrot-and-stick" approach to social upgrading. The case of the jeans cluster in Toritama (see Almeida's chapter) is an example of how to integrate the policy and "police" approaches. The public attorney started a crusade to make firms diminish the water pollution over the town's scarce water resources. He knew that such small firms would not have the capacity to develop the technology and invest in sewage treatment in a short period of time. Thus, instead of simply fining or closing down the firms, he decided to involve other public actors within and outside the cluster to help the firms to find a solution.

The difficulties of public organizations to integrate the police and policymaker role are sometimes embedded in their own mission or legal statement. Some law enforcement officials or regulators are not allowed to suggest solutions or help regulated firms to comply with the law (e.g., the state environmental agencies in Brazil). Officials or their organizations can only state whether or not a firm is complying with the law, and what are the consequences if they do not comply. There is no space for negotiation or a more constructive interaction between law enforcers and firms. Public officials would have to go beyond their formal duties to make the constructive interaction possible, but this may be against the law. The case in Toritama is peculiar, as the public attorneys in Brazil have acertain degree of flexibility to act in order to enforce the law over firms, including negotiation of deadlines and support from other organizations.

Policies to socially upgrade need a certain degree of flexibility to adapt to the diversity of situations and uncertainties faced by firms and clusters. They respond differently to distinct policies and require customized solutions to innovate and socially upgrade. The public organizations are more effective when they manage to have a flexible role and work together to provide specific solutions to firms, such as in the case of Toritama. The public attorney forced firms to work with universities and technical assistance agencies to find out a way to reduce water pollution. This interaction resulted in the development and adoption of a customized technology for effluent treatment. In the case of organic coffee production in Mexico (Damiani's chapter), buyers were flexible to wait and help producers to adapt their plantations to the quality and organic requirements.

Final Remarks

Globalization has created new opportunities for small firms in developing countries, as they can have easier access to new markets and information and benefit from the impacts of Foreign Direct Investment (FDI). Clusters and their firms need to acquire innovation capacity to upgrade and be able to take advantage from those opportunities. Many clusters in LDCs have had incentives to upgrade economically, as well as socially, as many opportunities also come with social demands. However, globalization also poses new challenges. Competition has increased and it is easier to move production from one place to another. This can lead to a "low road" race to the bottom with negative consequences to labor, communities and the environment, which may be sacrificed to maintain the price competitivenesss of clusters. Moreover, not all clusters and firms may be able to access the benefits of globalization and those which do access the benefits may remain in the hands of few, having limited trickle down spillover effects to spur local sustainable development.

This book is an attempt to understand how and why clusters and small firms have been able to develop innovation capacity and socially upgrade: formalizing, improving their environmental and labor standards and having a larger social impact in the communities they operate. The chapters in this book provide analyses that enable us to move the agenda on social upgrading, as there is no established comprehensive framework over this topic. This book tried to withdraw from the existing rich literature on small firm/cluster/cluster upgrading and connect it to other relevant bodies of research on other relevant topics for the understanding of social upgrading, such as global value chains, environmental/labor issues, social capital, innovation and local development. This helped us to reflect on some important points and questions over social upgrading, but there is still a long way to go in order to research and undertand how to make clusters and their small firms socially upgrade to bring sustainable development to localities and communities around them.

Index